WOMEN IN WAR

Shelley Saywell is a television researcher and producer whose previous credits include *The Ten Thousand Day War*, an acclaimed documentary series about the war in Vietnam. She and her husband live in Toronto.

Shelley Saywell

WOMEN IN WAR

Costello 1987

First published by Penguin Books Canada Limited, 1985
This edition published in Great Britain by
D.J. Costello (Publishers) Ltd., 1987

Printed in Canada by Gagné Printing

British Library Cataloguing in Publication Data

Saywell, Shelley
Women in war
1. War — Women's work — History — 20th century
I. Title
909.82 HQ1154

ISBN 0-7104-2033-1

CONTENTS

For Tony

ACKNOWLEDGEMENTS

In the three years it has taken to research and write this book, many people have given generously of their time and support. My gratitude is deeply felt. To Michael Maclear for his encouragement and faith in me throughout the years. To Debi Goodwin for her friendship and assistance, which included French translations. To Julie Smith-Eddy who kept up my spirits, and to all my friends who kept me from total hibernation.

I am indebted to Cynthia Good, editorial director of Penguin Books Canada, who discovered the project in its infancy and immediately offered her interest and guidance. I am equally indebted to my editor, David Kilgour, for his tireless patience, wisdom and above all enthusiasm.

Special thanks to all those people whose help has been so valuable: to Eugene Poznykov who arranged my visit to the Soviet Union, and to Ludmila Yenutina who was my guide there. To André Degon for his help with the interviews in France, and to both him and his wife Edna for their hospitality there. To Anne McCourt, photo research; Julia Schtivelman and Anna Malokina, Russian translations; Marianne Gilbert, Italian translations; Axelle Janczur, Spanish translations; and to Charles Fishman, Ephraim Ben-Mattityahu, Richard Hahn, Maria Elba, Laurence Tonnellier, Marilyn Edgerton-Mallard, Debby Littman, Jenny Baboolal, Jean Cottam and Anne Valery for their help and advice.

My gratitude to the Canada Council Explorations Program. Without their financial support this book would not have been possible.

My deepest gratitude is to my family. To my parents, whose love and support have been the foundation; to my sister Patricia who gave up a summer to help with research; and to my brother Jim who was my guide and interpreter in Italy. Most of all, to my husband, Tony Tobias, who convinced me I could write this book, and who has remained the most faithful and unending source of encouragement and inspiration.

Finally, I wish to thank all the women who shared their time and memories with me. They do not all appear by name in the pages of this book, but each has contributed to its spirit.

INTRODUCTION

In 1980 I was working on a television documentary about the war in Vietnam, and in the course of watching footage of American soldiers in action I saw a woman run across the screen. The image stayed with me. About the same time, I boarded a streetcar in Toronto one day and sat down beside a woman who, for inexplicable reasons, proceeded to tell me the story of her life — including four years in a concentration camp during World War II. In the weeks and months that followed, like someone who has learned a new word and hears it everywhere, I kept hearing of women who had served in war. I became fascinated with the idea of seeing war through women's eyes.

As I began to seek out these women and research the subject in libraries and archives, I was reminded that women have always been in battle. There is hardly a nation that does not have in its history at least one martial heroine. What astounded me was the extent to which women have faced combat in modern war — nursing, reporting, fighting, spying, bombing, killing, dying. Yet I could find little written, in their own words, about those experiences.

I thought it would be interesting to hear from women of different countries about the experiences of war they had in common. I also thought it important to hear not only from acknowledged heroines but also from unknown, ordinary women who were caught up in extraordinary events.

My first problem was to find a way to limit a limitless subject. Since I wanted first-hand accounts, World War II seemed a good place to begin. But war has occurred in virtually every part of the world since then, and choosing only a few for close examination was a difficult process. In the end I chose to deal only with "familiar" wars, the wars we in Europe and North

America have participated in or been directly affected by. This enabled me to keep historical detail to a minimum, focusing the attention on the personal story of each woman.

I travelled to the United States, Britain, Europe and the Soviet Union to interview women. I also found a surprising number of veterans in my own city, Toronto, which has a large immigrant population. It was here, for instance, that I met Ileana and Maria, the political exiles from El Salvador.

My interest was in the human element — how each woman had made her commitment to war, what the nature of her experience had been and how it had changed her. I had many questions about war seen from a female point of view — the ability to kill, the emotional toll of combat, the treatment of women combatants by their male counterparts.

The things I expected to hear and perhaps, in the beginning, wanted to hear were not always said. I wanted to hear that women are in every respect as capable as men: instead I heard about their weaknesses as well as their strengths. I expected to hear a great deal about male chauvinism, but I also heard about men who treated the women they fought beside as equals, as comrades. I wanted to hear that women are innately more pacifist than men, but I learned that they can be every bit as determined in their willingness to kill and die for their beliefs. At the same time I was often told that ultimately women are mothers and want above all to nurture and preserve life.

There were times when I felt overwhelmed by all the contradictions, but in war, more than any other human experience, contradictions are rampant. Rights and wrongs become blurred: human nature is at once at its most bestial, and most elevated. Even defining war itself was sometimes difficult. Official declarations of war often did not coincide with the actual combat experiences of the soldiers who fought in them.

There were things some of the women would not discuss, photographs they would not allow to be published. Sometimes this was a result of our different ages, cultures or moral values; sometimes it was a question of maintaining their anonymity or that of others whose lives were still in danger. Because these conversations are not published in interview format, the reader will not hear the questions that were skirted. But perhaps what was left unsaid can tell us something too. I hope I have succeeded in letting the women speak for themselves. I have provided only enough background detail to set each story in its historical context. Where I considered it relevant or necessary, I have pointed out which experiences were common, which rare.

I fell in love with these women. For the past three years I have listened to the most compelling events of their lives. Though I did not always share their beliefs or agree with the wars they fought, I could not help admiring the solitary courage they displayed. Many of them operated alone or as the only woman in a group of men. They combated not only fear of injury, as do men, but also fear of the unknown. They often had to combat a general abhorrence to the idea of women fighting or serving in any military capacity: it took courage to face being misunderstood, suspected of servicing men rather than serving a country or cause, and courage in coming home to people who could not or would not understand, who did not even recognize them as veterans.

In the end, however, none of the women interviewed for this book wanted to be thought heroic. They talked most of all about the waste and destruction war causes ordinary people. There was little that was glorious in the memories they shared.

If the stories told by women in this book say anything, it is that war is not merely the struggle between nations or armies. It is the sum of specific acts of murder and destruction, of cruelty and kindness, and of courage. Whether they are on the same side or not, no two people fight exactly the same war. For every war described here, millions more remain untold.

What these stories also prove is that in today's world men and women share the responsibility for the fighting of wars and the fight to preserve peace. War is not, can no longer be, a male domain.

Shelley Saywell
Toronto, January 1985

HOLDING THE LINE
Britain, 1939-45

And I am sure of this — that when victory is gained we shall show a temper as admirable as that which we displayed in the days of our mortal danger. In all this the women of Britain have borne, are bearing and will continue to bear a part which excited admiration among our allies, and which will be found to have definitely altered these social sex balances which years of convention have established.

Winston Churchill

When you landed . . . they'd be looking behind you as you got out of the plane, to see who the pilot was. They couldn't believe such a tiny girl could handle such a big, heavy airplane.

Diana Barnato Walker,
Pilot, Air Transport Auxiliary

Women were doing everything in England during the war. Churchill asked us to, remember. But I don't think the army girls were ever very popular. I think most people thought we were loose women, just there to entertain the troops.

Joan Cowey,
Anti-aircraft battery, Auxiliary Territorial Service

If you ask British women today when their emancipation really began, they will tell you that it was during the First World War, when women took upon themselves roles that were extraordinary in those times — not only running society at home but also going off to the trenches of Europe to serve as doctors, nurses and spies. When the Second World War erupted twenty-five years later, the example set by their mothers meant that young women could again choose from a variety of roles. As author and First World War veteran Vera Brittain wrote in her book *England's Hour* about the events of 1940, the war no longer claimed only soldiers' lives. It had become, amid the sandbags, barbed wire and smouldering ruins of England's cities, a war that did not distinguish its battlefields or victims. The crisis that faced Britain as she stood alone against Germany throughout 1940-41 necessitated freeing up all able men for active service. The running of the country's war industries and agricultural production fell into women's hands. By 1941, however, military shortages were so severe that Prime Minister Churchill asked for the conscription of all single women between eighteen and thirty years of age into the Auxiliary Territorial Service (ATS) — the women's auxiliary of the army. Their jobs were not only in offices, but on searchlight and anti-aircraft sites, driving trucks, filling sandbags and digging roads. They were toughly trained and lived in rough conditions, either in long Nissen (metal) huts that scarcely kept out the snow in the winter, or under canvas. Despite the fact that women had served in World War I, the public was not entirely liberated in its attitudes towards women at war in 1940, and army women suffered from both animosity and, later, anonymity. One former ATS member wrote, "You could say that the forgotten army was not the one in Burma, but the one in skirts."[1] Her sentiments are shared by many army women, although some would argue that they spent more time in battledress than in skirts.

Women also joined the auxiliaries of the navy and air force. Women pilots flew for their country not in the air force but in a civilian organization, the Air Transport Auxiliary (ATA), which ferried war planes from factories to the RAF bases throughout Britain and later to the continent.

There were casualties in all these occupations, and studies on British women at war now refer to many of their roles as "quasi-combat," suggesting that the definition of combat duty sometimes grows hazy. Embodying this ambiguity were the women spies, trained as civilians to avoid the combat issue altogether, who worked for the Special Operations Executive (SOE) and were dropped into occupied France and Poland. An unknown number of women agents worked in army, navy and air force intelligence. They served behind the lines in other war arenas, including Asia and the Far East. Because as women they could not be classified "combatant," and because they were signatories to the Official Secrets Act, their stories remain unknown.

Diana Barnato Walker flew Spitfires and Mosquitos through the densely crowded air space over Britain from 1941 to the end of the war. A beautiful woman still, in a wartime photograph she appears as an almost wistful aristocratic beauty. Today she lives alone in her country home in Surrey. In her home, sitting in front of a small November fire, she talked to me for hours about her personal experiences and also about other women who flew during the war. She still seems to appreciate the novelty and adventure of those experiences, but there is another side to this woman: the war for Diana was most of all an unhappy time. She lost many friends and suffered a deep personal loss with the death of her husband only a short while after VE day. Today she says, "When you leave I shall stop thinking about those times, because it makes me so sad. I'll just block it out of my mind."

Joan Savage Cowey served on anti-aircraft gun-sites throughout the British Isles and then in Belgium. A short, solidly built woman with an attractive face and sparkling blue eyes, she still lives in a small square house that looks like all the other small square houses on her street in a small Canadian community, built for veterans, which she came to as a war bride forty years ago. Today she has eight children, most of them grown, and keeps busy with Canadian Legion (veterans') events. She has not let go of her ties with the Legion, despite the fact that most of the members are men, because she claims to identify more with them than with anyone else. She is a woman who enjoys talking about the war, specifically about what women in Britain did, because she feels people underestimate their contribution. She was angry at times when I spoke to her, recalling the sexist attitudes directed towards her after she had spent five years as a soldier. But she also showed a softer side, compassion, as she remembered the orphaned children and bombed-out cities of Belgium. Joan Cowey still thinks about those days all the time. She says, "After all, it's all we have. Do you know what I mean? It is really all we have.

"It's history now and people say, 'Fancy! In those days before women's lib, way back forty years ago, women flying Spitfires and Mosquitos? How did you get round to doing it?' But then nobody paid any attention. Everyone was just doing his job to help the war effort. Nobody thought about anyone else's job, and when the war ended everyone was so weary, everyone had lost loved ones, nobody stood around congratulating anybody else."

Diana Barnato Walker started to fly for recreation shortly before the war, to escape "life almost in purdah." The eighteen-year-old daughter of a millionaire racing car driver had a strong will and an independent streak. She was attracted to the sport that offered a sense of adventure and escape. "You were guarded in those days. You had nannies and then you had governesses; you couldn't go anywhere by yourself. When you went to parties you had a chaperone. Even at eighteen you were clung to, and I wanted to get away. So I thought, 'Well, if I learn to fly, nannies and governesses won't come up in an airplane.' That was the basic underlying reason.

"Just before my first solo, I was taxiing out by myself when a man yelled and rushed over to the side of the cockpit. He was terribly badly burnt. I had never seen anything like that in my life. And he put his hand, which was horribly disfigured, on the side of the cockpit, and his face very close to mine — I was only a young girl who never saw things like that — and he said, 'Don't fly, Miss Barnato. Look what it has done to me.' A rather sobering introduction. But I went up anyway and managed to get a grand total of ten flying hours before the outbreak of the war."

When war was declared in September 1939, Diana signed up as an ambulance driver for the Red Cross, not considering that there might be a job for her as a pilot. The war came to Britain with a strange slowness. As Poland fell and France and the

Low Countries were invaded, British citizens watched the approaching storm with a sense of unreality. The atmosphere at government and military headquarters, however, was tense. Preparations were made for a drawn-out war and the possibility of invasion. In support of the Royal Air Force a civilian organization was established, the brainchild of a BOAC director, Gerard d'Erlanger. He foresaw the need for pilots to aid the RAF in ferrying the enormous numbers of fighter planes that might be needed, from factories to airfields across the Isles. The organization he founded was named the Air Transport Auxiliary (ATA), and for it he selected one hundred experienced pilots who did not qualify for the RAF. These pilots included World War I veterans, slightly disabled male pilots and eight female pilots.

It was a rough time for these eight experienced women pilots. The press got hold of the story and revelled in its "drama." The English were still living in the false security of the "phoney war." They had been geared up for bomb attacks, but none had come. The threat of war had not yet become a personal one for most of the British, and they were more concerned about high unemployment than about national defence. Women pilots were often seen as frivolous wealthy women who were stealing men's jobs. Diana recalls, "They had a tough time because they were proving themselves. And if those women hadn't stuck with what they were doing, the likes of me who came in nine months later wouldn't have had such an easy ride. They were proving themselves to the men, to show they could do it. There was no resentment from male pilots; they just weren't convinced that the women could do it." The resentment came from the public. In one scathing editorial, a journalist wrote: "We quite agree that there are millions of women in the country who could do useful jobs in the war. But the trouble is that so many of them insist on doing jobs which they are quite incapable of doing. The menace is the woman who thinks that she ought to be flying a high-speed bomber when she really has not the intelligence to scrub the floor of a hospital properly, or who wants to nose around an Airraid Warden and yet can't cook her husband's dinner."[2]

Part of the problem was that the press ran exaggerated accounts of the women pilots' salaries and duties. In fact the women were, and continued to be, paid less than male pilots. Furthermore, in the beginning they were not allowed to fly any operational aircraft, but every time another article appeared with pictures of the "girls in bomber pilot togs," their critics were vocal. "I am absolutely disgusted," one woman wrote in response. "When will the RAF realize that all the good work they are doing is being spoiled by this contemptible lot of women?"[3]

It was not just their unpopularity that made the first year difficult for the ATA women. "They were only allowed to fly Tigermoths at first," recalls Diana. "They are open aircraft, and their missions were to fly all the way up to Scotland, to Prestwick, which meant four hours or so of flying with a refuelling stop, and they used to come back in the night trains, sitting on their parachutes in the unheated corridors, get into Paddington and out to Hatfield by train, or hitch-hike, and do the same thing the next day. One of them, Amy Johnson, was killed returning from a Scotland mission. It was very hard for them, but eventually they were allowed to fly over 130 different types of aircraft. When the phoney war ended, so did the resentment, because when the real crunch came we were desperately short of pilots, and suddenly everyone was needed."

On 10 May 1940 Prime Minister Winston Churchill took office. A man passionately committed to the fight against Nazism, he believed that this was a moral fight against evil, worth great sacrifice. The day Hitler invaded France and the Low Countries, launching his war against western Europe, Churchill ordered the RAF to drop their first bomb on Germany. The phoney-war mentality evaporated overnight. Only seventeen days later France was crumbling, and allied forces at Dunkirk called for air protection to begin an evacuation. Weather played friend and foe; the RAF lost over a thousand planes and their crews in support of the naval rescue missions that eventually saved over 300,000 men (and some army women). Over 558,000 allied troops were eventually rescued from Europe and taken to Great Britain.

By late June France and the Low Countries had fallen. Marshal Pétain of France signed the armistice with Germany on 22

June in Paris. The European war had now become a German-British conflict. Prime Minister Churchill spoke of possible invasion and the call to sacrifice. Overnight the British Isles had become the front line, and in August the first Luftwaffe air attacks, in what would be known as the Battle of Britain, began.

The British pulled together, their baptism under fire producing a powerful collective anger rather than the cowering Hitler had hoped for. Despite the endless nightly pounding, the attacks were less destructive than had been anticipated. The RAF reported fewer losses than the Luftwaffe, and the Battle of Britain, in not producing England's surrender, resulted in her victory.

By the eve of the Battle of Britain in early September, British factories were producing planes at the fantastic rate of almost five hundred a month. The RAF needed considerable help in getting these new, minimally tested planes from factories to bases. The ATA expanded to a force of eight hundred pilots and, on the strength of the performance of the original eight female recruits, opened its doors to all qualified women. In total nearly one hundred women flew for the civilian organization. Diana Barnato Walker took the ATA test despite her limited air time: "My parents hated the idea. They thought I'd break my neck. I didn't think I'd ever pass the flying test with only ten hours' experience, but two of my RAF admirers talked me into it. They sat me down on the sofa and said, 'Pretend this is a cockpit.' In effect, I got the additional time I needed sitting on a sofa in my living room. I passed the ATA test."

Diana was accepted, but the same week suffered a fall from a horse that put her in the hospital. It was not until November 1941, at twenty-two years of age, that she officially signed in at ATA headquarters at White Waltham.

"My first impression of the ATA was that it was an extraordinary organization. It was Pops d'Erlanger's vision of collecting people together, so many varied types of people including women, and men who in the main weren't able to be in the services. We had two one-armed pilots, for example. They collected millionaires, butchers and bakers, young girls and First World War pilots — all sorts of people of all shapes and sizes, and the only thing we all had in common was that at some time or another we'd learned to fly. In addition we had girls who came over from Canada, the United States, Australia and South Africa, all these places where they were trained pilots and wanted to do something and had heard about the ATA. D'Erlanger moulded it into this extraordinary little group which moved a fantastic number of airplanes."

One Canadian pilot who joined the ATA recalled her first impression of living in wartime England and working for the ATA: "We were billeted with English families. I lived with a woman who worked in our cafeteria. Whatever they asked us to pay them for board we would. Some places the food wasn't too good, but they didn't have much food at all. We were given ration coupons, and we used to give our extra ones to them. Whenever I got a parcel they'd be hovering over me, hoping, I suppose, for a few extra cigarettes or something. I never saw an egg or an orange the whole time I was there. Other times we got stuck overnight at an air force base. They had Nissen huts, long metal huts with no place to cook or anything. What used to get me was that in the winter they just had a teeny weeny stove in them, and this was supposed to keep you warm in the night. Good grief! Those Nissen huts were real cold, just a metal shell, a cement floor. Some of the living conditions were dismal — but who cared about that? It didn't bother me."[4]

Diana: "I lived at home for a while and then got posted to another aerodrome [at Hamble] and was billeted with a family nearby. I rode my bicycle into work in the morning. We were a civilian organization, but we did have officer's status, a uniform which included flying boots, leather coats and trousers, and lots of rules and regulations. Still, we never had to salute or drill.

"By the time I joined they had formed an ATA flight school where one got lessons on fifteen different kinds of planes. I was eventually trained as a Class IV pilot, meaning I could fly the heavy twin-engine craft. A few women were qualified to fly four-engine planes. We learned about things like hydraulic systems, retractable undercarriages, different kinds of propellers and so on. We also had handbooks which showed us how to fly any variations of these fifteen basic types, simply by following good, straightforward instructions.

"We reported in to work at nine o'clock every morning and saw the operations officer. He or she would have already rung up RAF central command to find out what planes needed to be moved and where. He would give us our chits [orders for the day], and then we'd go to the weather officer to get the local conditions. That is essential in British weather."

Sometimes, in really bleak weather, the pilots would spend the better part of the day waiting for the clouds to lift. The women would play cards, knit, drink tea. When the conditions were right the women flew a number of missions each day. It was not a nine-to-five job, and there was no guarantee that the pilots would not be forced to sleep out at a distant base if they arrived late or if the weather was bad. They generally returned to their base in a taxi plane piloted in turns by them all, usually one of the larger Ansons or Fairchild-24s, which would collect them throughout the country.

The women flew their missions without radio contact. "Our navigation was done entirely by map and compass readings, and dead reckoning — just looking at the ground," says Diana. "Part of the reason for this is that the aircraft were so new that many did not have radios in them yet. But the main reason was that they didn't want lots of people chattering on the radio, interfering with the very busy channels of the RAF communications. So we were taught to contact-fly, and were supposed to stay out of the clouds." Contact flying is simply navigating by flying low and keeping the ground in sight. With the use of maps pilots could spot key landmarks and navigate without help from ground control. Because English weather conditions could suddenly change for the worse, bringing rain and fog, the pilots often lost sight of the ground. Then, in what they refer to as blind flying, they could navigate with only the plane instruments to indicate direction, altitude and speed. Diana learned this skill as she went: "People like me, who only had ten hours' flying before the war, did not know how to blind-fly. It may seem unbelievable to think you could fly all that much and not know how to fly in a cloud, but people like me didn't know. I found out by trial and error and frightening myself. I think the whole thing was a mistake. I think a lot less grey hairs would have been grown and a lot less people would have killed themselves, or badly frightened themselves, if we'd flown with radios.

"I remember one day before I could instrument-fly, I was flying on a lovely summer day, everything was all right, and suddenly I was in cloud, just like that. It was my first time in a cloud. I did all the things I had been told, but I was very close to the ground and slightly off course. I'd just come off leave and was still wearing a skirt and stockings. I thought, 'I can't bail out because I'll look so silly coming down in a skirt with my stockings and panties showing.' It was so stupid; I should have bailed out. I broke at treetop level and thought this was too low. It was pouring with rain. Thank God I saw a little plane landed on a grass airfield that was covered in water. It was not a good idea to land in water in a Spitfire because the flaps would get caught and break and the plane could nose over, but I had no choice. I landed it all right and taxied in. It was teeming with rain. I shut off just near a Nissen hut, and a very tall man came out of the hut with a huge camouflaged cape over his head and walked over to the side of the airplane, putting up the cape to keep me dry. I got out of the cockpit and my knees collapsed, I was so frightened. I couldn't let him know that, so I pretended I had kneeled down to get my bag out of the plane. When we went into the hut I stole a glance at the bulletin board to see where I was — I didn't want to ask."

It turned out that Diana had landed at the base of a blind-flying instructor. When he learned that she had no idea how to blind-fly he gave her lessons. But in general the ferry pilots were supposed to avoid clouds where possible, even if they had blind-flying experience. Diana explains: "Another reason for trying to avoid clouds was that we had to be identified by the gunners on the ground. We did get shot at sometimes. All the people on the anti-aircraft sites were marvelous, and they were trained to know what every aircraft looked like. But if we were flying across the Bristol Channel, which is very misty and murky, in a Hurricane, we could hear the shots and see the little black tracers coming at us. One day there was a notice on the board that anybody who'd been shot at over the Bristol Channel should report it. A lot of us did, and after that things improved.

The anti-aircraft people couldn't tell if it was a Hurricane or a Messerschmitt, I guess, in all that fog."

On the ground below, a young anti-aircraft crew member, Joan Savage Cowey, adjusted her heightfinder and scanned the skies: "We had hours of aircraft recognition lessons, for fear we'd shoot down one of our own. That always bothered my mother, because my brother was an RAF pilot. She always said, 'Be sure you don't shoot down the wrong plane!'"

Joan had quit school to do something in the war. The tragedy at Dunkirk had touched her personally: "A lot of the boys I went to school with were killed at Dunkirk. Half of the village of Henfield, my village, didn't come back. You want to know what that does to a small village? It made us mad, that's what.

"I was a bit too young, but I remember I wanted to get into a uniform of some kind. My father was a Royal Artillery man, and I always wanted to do that. After Dunkirk, you could say I wanted to do it even more."

At eighteen years of age, the five-foot two-inch daughter of a chauffeur showed her independence and a "stubborn streak" and despite protests from her parents signed up to join the army. "My parents thought it was a terrible thing for a young girl to go into uniform. They thought women should be with the Red Cross rolling bandages or something. I couldn't see myself doing that. I think they thought women's place was in the home, and there was other war work that we could do. But to join the army and go with the men — that was bloody awful.

"The whole idea appealed to me. I would have loved to be a guerrilla fighter. I always read about them. I admired the Polish and Yugoslavian girls, and I thought, 'Aren't they lucky that they can fight.' The whole thing appealed to me, the training, the adventure. Of course I was just a teenager. It's different when you are a teenager — you never consider the dangers. So I signed up and went down to Guildford."

If the public had a hard job accepting the first female pilots, they showed no more understanding towards women in the army. Joan's parents reflected the views of many people of the day. The women's Auxiliary Territorial Service, the ATS, was the largest and least popular of the women's services. The women in the army underwent strict military training and had a reputation for being tough. When the service acquired full military status in 1941, they were subjected to drill, physical training and military discipline, including court martial. They were often stationed in camps recently vacated by men, and in conditions that were considered "unsuitable" for women. Such conditions included latrines without doors and camps without barracks, where the women, ten to a tent, were kept awake at night by the rustling of armies of rats. Unlike the other women's services — the navy, air force and civilian — the ATS had little glamour. Even the uniforms were dull by comparison, drab khaki battledress in contrast with the blue uniforms of the Women's Auxiliary Air Force (WAAF) and the Women's Royal Naval Service (WRNS, or "Wrens"). Wrote one critic to *The Times,* "Khaki is a colour detested by every woman and makes a well-developed girl look vulgar."[5]

There were 450,000 women in uniform in England by the height of the war, more than 200,000 of them in the ATS. Of these, approximately one-quarter served on anti-aircraft (AA) sites. In October 1941 the army agreed to experiment with the first mixed male-female AA site, in which women did everything but actually fire the guns. It was a controversial move, born of the necessity to free up more men for the front, and it did nothing to help the reputation of army women. The first mixed units were scrutinized by press and public. Men and women's living together sparked waves of hostility. Joan remembers, "There was the feeling that we were sort of loose women living in tents with men. They called us 'officers' groundsheets' — we got that all the time. They didn't really know what intense training we went through; they just thought we were there to entertain the troops. The American soldiers were worse. They'd say the Brits had it really good, having us girls along to keep them happy. The whole thing — all the criticism of us — just had to do with sex."

Rumours of immorality coupled with the hardships of army life did nothing to help recruiting. But the necessity of attracting more women into the service was becoming critical. The ATS women would eventually serve in eighty different specialties, including the driving of three-ton trucks and tanks and heavy work such as road building. These previously all-male domains were now being turned over to the ATS. Churchill's daughter was an officer in the ATS, so he had a personal interest in seeing that the organization got the soldiers it needed. In December 1941 he introduced a bill in Parliament calling for the conscription of all single women between twenty and thirty years of age. Those who chose the services were to be directed into the ATS without choice. The other two services were so much more popular than the army that they had waiting lists.

It was a controversial move that elicited immediate and heated debate. Declared one Member of Parliament, "We are the first civilized country to propose that women should be drafted. The Nazis tried it and failed. How can we hope to succeed where the Nazis have failed?"[6] But the argument really centred on the issue of morality in the ATS. According to Member of Parliament Stewart, "There is no doubt that the ATS has a thoroughly bad reputation. The conditions of the camps are bad, the physical condition of some of the girls is bad, and the whole service has a bad name. The War Office must take notice of the stories that are being circulated. There is a general impression that the ATS is not the sort of service that a nice girl goes into."[7]

The War Office did some checking and fought back with statistics; the pregnancy rate of unmarried women in the services was actually lower than in civilian life. In one camp, rebutted the Secretary of State for the War, "We found not a single case of venereal disease, and only one woman was pregnant. She had been married for five years. (Laughter) I hope that the honourable members will do their best to put a stop to this slanderous campaign. (hear, hear!)"[8]

Of the three women's services the ATS did have the highest pregnancy rate in 1941, and the figures would double by 1943, but it was lower than in civilian life. Illegitimate pregnancies were escalating rapidly in all sectors during the war. With up to a million foreign troops on British soil at any given time over the years, the increase was not surprising. One of every eight children born during the war in Britain was born to a single mother.

The conscription bill was passed that December, and the service took on a lot of drafted and miserable women. The first thing a new recruit encountered was the medical. Said one former ATS woman, "The medical was hilarious. We were stripped and prodded and told there was no time for niceties. And then it was into hygienic for de-lousing."[9]

Barracks were assigned and uniforms issued. The latter consisted of khaki battledress, a blouse and slacks, anklets, webbing or leather, brown lace-up boots with steel studs and tips, tin hat, leather jerkin, respirator strapped to the chest, water bottle to be slung on the hip and an anti-gas mask. Even the underwear was khaki coloured and popularly known as passion killers. They had been ordered to save the ordinance officer the indignities of dealing with feminine underwear. General Gwynne-Vaughan of the ATS suggested that, "If they like to provide crêpe de Chine undies of their own and wash them in the off-duty time, I see no objection and it would save the public purse." But she was relieved when the army-issued set was tried out by the Princess Royal, who "desired a set sent to her" for her inspection. Her verdict was that she would be willing to wear them herself.[10]

Basic training was hell, Joan remembers. For four long weeks the days started at six-thirty in the morning with breakfast, then inspection, then drill, lunch, lectures and "fatigues," which included scrubbing floors and washing dishes in the mess. And for the first time in their lives, women experienced army discipline.

"Everything had to be done the right way, the correct way, and it didn't matter how nice your bed looked, they'd come in for inspection, kick it and the officer would say, 'Now do it again,'" grumbles Joan. "I remember they called a parade on an awful wet morning, their favourite time to call them. We had studs on the bottom of our boots which we had to polish. A colonel put his stick across the back of my legs and I had to put my leg up to show my boots. He said they were dirty and I was told to get them shined. That's when you feel like turning around and walking. . . .

"The sergeants were terrible. They had trained in India and all these places. They were all sergeant-majors, and the roar of them! We almost needed earplugs. I saw girls crying, but there was no way they were going to get me like that. One sergeant came in the room and said, 'Now we'll see who's tough.' At six-thirty in the morning he'd come and bellow in the door, 'Up and at it.' We'd have to get dressed, and we'd run and run and march long routes, with packs on our backs and blisters on our heels. We didn't dare say we had blisters. We just came back to the hut, gingerly took off our boots and hoped the sergeant didn't see us wince."

Another ATS woman remembers, "We were taught by a lovely man who had only taught troops; up until that time he'd been on the northwest frontier. We made his life a misery. We were always falling about laughing, and he didn't take easily to being put down. He was always telling us how we weren't really women at all, in our dreadful uniforms with our toughness. A real woman wore a picture hat and chiffon!"[11]

"A lot of girls were miserable," recalls Joan Cowey. "The girls who were drafted had a hard time accepting it. I remember girls getting really hysterical in the barracks rooms — they couldn't cope. Some of them had never been away from home before, and they didn't think they could stand it. We couldn't answer back to anyone, and I'd see girls cry and cry. The ones from the poor backgrounds got really homesick because they'd never been to boarding schools or away from home before. But I loved it. I'd never let a man see me down. No man was going to get me down that low. I was a volunteer, and the volunteers kind of stuck together, but the discipline got to all of us. We were constantly bellowed at and for the smallest thing put on peeling potatoes for a week or something."

"I loved it," an ATS driver of a three-ton truck recalled. "People said women hated to be in uniform, that they missed their party dresses. Balls! It was lovely, all that way from home and boarding schools, with no one knowing who you were. The uniform took away your identity and gave you freedom — even if it was just to do something like have a pint of beer in a pub.

"There was some deep resentment between girls, especially of different classes," she continued. "It did lead to quite a few punch-ups between the girls, just to get acquainted. Some of them really hated the army. They'd get drunk and say, 'I cannot stand it anymore.' Then they'd go across the barrack square and holler up to the men's quarters, 'Paragraph 11!' (Everyone knew that was army regulations para 11 for compassionate discharge for pregnant women.) They'd be hoisted in, get pregnant, get out of the army and go have an abortion. Very basic, the army."[12] Eventually the army discouraged this by discharging pregnant women only after they were six months pregnant.

England was still very socially stratified in the 1940s, but the war was breaking down some of the class barriers. In the ATS barracks, rich, poor and middle-class women learned to live together. Joan: "We were a mixed bag, really. We'd have girls with titles, and nobody ever asked them what they were and nobody cared. It didn't mean much to me, my father was a chauffeur for a rich family. I got more fun out of the cockneys. They were so funny. Actually, the girls from the middle class got on the best. They'd been to boarding schools from an early age and were used to living away from home."

It was the same in the ATA. Diana Barnato Walker had never before met people outside the upper class on a social level. "We met a lot of people we wouldn't have met otherwise. In those days England was very stratified. One only met people from one's own type of upbringing. Here we were all doing the same thing, without distinction. It was wonderful." Another ATA woman pilot saw it differently. "I was a working-class girl who learned to fly to make a living. There was a clique of rich girls who had all learned because it was the 'in' thing to do. They owned their own planes, had their own friends, and you were never invited to their homes or parties."

After basic training Joan Cowey volunteered for anti-aircraft and was sent to firing camp in North Wales. The ATS was now experimenting with mixed units in which women did everything on the gun-sites but fire the guns. The experiment constituted a loosening of the definition of non-combat status. (The male gunners had combat status.) When the first recommendation for using women personnel on gun-sites was made in 1940, it could not be authorized under the existing definition of combat in the Royal Warrant of 1938. When the ATS received full military status in the spring of 1941, the Secretary of State for the War explained to the House that this would include duties on searchlight stations and AA gun-sites. The ATS "gunner girls" continued to make only two-thirds the salary of corresponding males because as "non-combatants" they were of less value to the state. The issue of combat status was side-stepped in a Defence Regulation that came out that spring, which stated that the Army Act applied to these women "in such manner, to such extent and subject to such adaptations and modifications as the Army Council deems necessary." An amendment to the Conscription Bill that was passed seven months later underlined the ambiguity of women's status by stating: "No woman should be liable to make use of a lethal weapon, unless she signifies in writing her willingness to undertake such service." All members of searchlight and AA batteries were, therefore, volunteers.

This was the most experimental and perhaps the most controversial role for women in the British army. It was to one of these gun-sites that Joan went after her passing-out parade. "It was forty years ago, and ours was an experimental group. I think it proved all right. The men seemed happy with us.

"After basic training I took my exams, passed and was sent on to the Royal Artillery camp, and of course by now I was thrilled to bits — I was going to get somewhere. It was all firing from morning to night, learning the instruments to fire the guns, determining the height and distance. They had planes going back and forth with sleeves or tow-targets on the back of them. There would be two hundred guns going off at once. It was all women and men all the way along the cliff. It was cold; I remember my hands would always freeze to the metal.

"The camp was strict, terribly strict. It's the best one in England, I've since learned. We did everything with the men there, on the parade grounds, in the classrooms, and they acted fine towards us. They accepted us. I've never heard anyone say otherwise."

The gun-site to which Joan was eventually posted consisted of long Nissen huts, each housing eighteen women. The windows were all blackened. The men's barracks were situated at the furthermost point in the camp. There was a "Naafi," or social club, where the men and women could drink tea, play the piano, sing and dance, a mess, the command posts and gun-sites.

There were four guns on each site, heavy 4.5 static guns and the mobile 3.7 AA-guns, all painted army-olive green. These were fired by men only. The guns were surrounded by sandbags, and operations such as instrument readings were carried on from behind a low concrete emplacement, just opposite the gun-pit, where the heightfinder and predicter stood. The women peered through telescopes and barked out distances and co-ordinates to the gunners. "We worked outside, just behind the guns," says Joan. "We spent long shifts outside at night. As soon as the spotter would shout 'Plane!' the alarm would go off. It was my job to find the plane with my telescopic heightfinder, and keep the plane in the centre of the viewfinder. These were mathematical instruments designed to judge distance and height, and with those co-ordinates the gunners could line up the guns." Below the gunners and women on instruments, in a dug-out, was the command post with its telephonists, who were in constant communication with other sites in the area.

Reaction to the mixed gun-sites proved both curious and comical. One "ack-ack girl" wrote in her diary, "Whenever a mixed battery took up site . . . its perimeter was haunted for the first few days by local people staring at the female curiosities and hoping, perhaps, to catch a glimpse of a pair of khaki bloomers dangling from the barrel of an ack-ack gun."[13]

Dignitaries and reporters descended on the camps to inspect the experiment firsthand. Churchill, perhaps in reference to his daughter's role in the ATS, commented, "I feel just like an anxious parent inspecting a preparatory school."[14] A *Times* correspondent wrote, "First evidence of one of the most far-reaching experiments ever made in the British army is meeting with great success. They [the women] have shown that short of actually manning the heavy guns, they are the equal of men."[15]

In 1942 Germany launched air attacks against British cities that were not primary industrial targets but rather of cultural and architectural importance. Nicknamed "Baedeker raids," they were in reprisal for an RAF attack on Lübeck. More gun-sites went into action, and the ATA ferried a hundred thousand planes to RAF bases throughout the Isles. The pilots worked long hours and got only two days off in every ten. But they could not take a break from the war, and days off were often spent in air raid shelters. One woman pilot told the author: "Sometimes on days off I'd go into London to visit some friends. The underground was terrible during the bombing warnings. It was just terrible down there, everybody crowded and you could hardly breathe. You had to stay down there till the all-clear sounded. Then I'd go for tea to my friends' house. They had a dining-room with a great big square metal table. I think every time I visited them I ended up crawling under that table, putting up wire around it, in case the house got hit during the raid.

"On days off we used to visit the wounded soldiers in the hospitals. A lot of the boys didn't have anyone to visit them and they got used to us coming. There was one guy we used to go to see, who was all bandaged up. He had landed on a runway and collided with another plane, and both exploded. He was very badly burnt. He just had a slit for breathing and he got very excited one day because his bandages were supposed to come off. He said he wanted us to be the first to see him. The nurse took the bandages off and I almost passed out. I hung on to the bed, went over to the wall, and felt my way all along the wall till I got outside the room. Because when she had taken the bandages off he didn't have any eyelids or eyebrows, and he was staring right at us. It made me sick to my stomach. We'd see kids with one leg or one arm, and they were really depressed. That was the reality of the war, for me."[16]

Diana: "I always thought, if I crashed, I'd rather be killed than disabled. A damaged woman has nothing. Men can function without a leg or arm, or burnt, and they are considered heroes. People accept them. But a damaged woman? I think it was my greatest fear.

"We were non-combatant in our classification, and that never changed. We sometimes flew aircraft with the guns loaded. The ammunition was in them, but the guns weren't cocked. We couldn't press a button to shoot anybody, and yet we were there to be shot at. It was a fine dividing line."

"Sometimes when I was flying a loaded plane, I felt like turning it round, towards Germany, and letting them have it," recalls one woman pilot. "The war made me so angry. I would fly across the countryside and see it all blown up, things destroyed, and it made me furious to think that they could do that. But of course, if I'd ever gone off and dropped a bomb myself, I would have been kicked out."[17]

"We were there to be shot at," says Diana. "One time I was flying with Jim Mollison, a very famous man who had flown all kinds of records in America. Jim was piloting an Anson, and a whole bunch of us were in it being taxied back to White Waltham. I was sitting up in the front with Jim. Suddenly we saw a dark cloud with the sun shining down on it and an aircraft darting out of this cloud, coming straight toward us. Both Jim and I thought it was a Mosquito, flown by an ATA pilot. We thought he must be on instruments, because he didn't put his head up and hadn't acknowledged us. Then somebody saw tracer bullets coming out against the dark clouds and Jim said, "Jeez, it's a Gerry!" He yanked the throttle and put the plane back up into the overcast, and we hid in the cloud until the Messerschmitt went by underneath and I saw the swastika cross marking on it. Then I remembered that as we'd flown over Reading there was smoke coming up from the railyards, which I had thought was smoke from the train engines. We found out later that it had been bombed by this Messerschmitt.

"We got back to the aerodrome and everybody was scurrying around with tin hats on saying, 'Did you see the Gerry?' We said, 'Yes, we got shot at.' Jim said, 'Come on Diana, let's go have a cup of tea.' So we went into the mess and had a cup and he said, 'You know, by tomorrow rumour will have it that I shot him down.' And lo and behold about a week later the press got a hold of it and wrote dramatic stories! It would have been a good bag for the Germans if he had got us — all those pilots in one shot." After this incident, the ATA thought of assigning a gunner to each taxi Anson, as they had done in the early part of the war, but in the end it was not done.

Diana: "Most of our accidents and deaths were caused by the newness of the craft we were flying. They were coming right out of the factories, untried. They'd just been pushed round the circuit for five minutes by the test pilot — because of fuel economy and the magnitude of plane production. So the ATA pilot in effect became the test pilot, since we were the first ones to really fly the plane. We had a 'snag sheet' on the back of our delivery chits, and we wrote down whatever was wrong. Things did go wrong. Luckily the ATA had given us such good training we could handle most problems. We knew what to expect and how to avoid things before they happened. We also had to beware of the barrage balloons that were planted all around the important factories.'' These were huge silver hydrogen-filled balloons that looked like small Zeppelins, attached to the ground by steel cables. They were grouped to form a protective covering over strategic spots, making it impossible for enemy craft to get close without getting caught in them. All the aircraft factories had them, adding to the ATA pilots' daily challenges. The steel cables were controlled from the ground. If a pilot wanted to land at an airfield through the balloons, the controllers would clear a 'lane' for him or her.

''Lots of us did fly into the balloons, especially in fog or bad weather, which England is famous for.'' Another woman pilot told the author: ''I got caught in the barrage balloons once. I was out in a Defiant, and I was told to run left and instead I turned right because the plane had begun to heat up on me, and I had to get her back down. I had completely forgotten about the balloons. I flew into the barrage, got caught but just on the edge of it, fortunately . . . though I must say I got a little uptight for a few minutes there. But we really never were very scared. It never dawned on me when I joined up that I would be flying where they were fighting, or that we could get shot down as easily as anybody else. It just never crossed our minds.''[18]

Diana: ''I don't think I was scared. You're not scared when you are young, anyhow. You don't know what you are in for. I'd be scared now, of course. You weren't scared until afterwards, because your training was so good that you did all the right things in sequence, and in my case got away with it.

''I remember a girl named Diana Ramsey who took off from Langly in a typhoon or a tempest. Something went wrong, her throttle got stuck, and so she flew a few minutes towards our aerodrome at White Waltham. We watched as she did a circuit round, and I suddenly heard her engine stop, and thought 'My god, what has happened?' She was trying to land, but of course she was going much too fast, she had no time so she had switched her engine off to try and lose some speed. She hadn't lost enough speed. . . . There wasn't enough time to judge correctly. She came straight across the White Waltham aerodrome, flew over the church just missing it by a few feet, and we rushed over towards the scene of the crash. We ran across the field and came to two enormous oak trees whose branches had been torn off. On the ground were gouges where her propeller had cut into the earth. We rushed towards a ditch and the woods, and saw the tail of the plane. Then we ran past the left wing. I was praying for her, 'Poor Diana.' We came across a little clearing where all the trees were broken, it looked as if a dinosaur had had a fight in there, and there was the rest of the fuselage, with one wing and the engine. I started looking on the ground, thinking Diana must be on the ground somewhere, dead, and suddenly I looked up and there she was sitting astride the remnants of the fuselage, trying to reach up into the tree where her hair ribbon was hanging. The guy I was with noticed there was gasoline all over the ground and yelled, 'Diana, Diana, come down, it might blow up!' So Diana came down rather reluctantly, perfectly composed, as if she'd just had a perfect landing. She walked along with us, quite all right, until we got out into the field where some cows were grazing. Suddenly she turned ghastly white and stopped dead in her tracks. She turned round to me and said, 'I can't, I can't . . . I'm frightened of the cows!' So the fearless airwoman became completely feminine again!

"That has always been one of my favourite stories. But fifteen ATA women were killed flying during the war. One was killed in a Spitfire, and another had something wrong with her plane and crashed trying to land it. Anna, a very sweet girl, got caught out in an Oxford in terrible weather. . . .

"I thought about it sadly for about five minutes, literally. And then threw it away. You couldn't keep it with you, because you would have been confounded. Everyone was getting bumped off, your friends, your family. You were losing the people that you loved, and if you thought about it too much you couldn't have survived. When anyone in the ATA lost somebody, they would be moved to another ferry pool to take their mind off it. Our commanders were very sensitive that way."

The deaths of ATA pilots, 173 over the war years, were mourned in full formality, the most "military" manifestation the civilian organization allowed itself.

"All of this was taking its toll," recalls Diana. "Losing loved ones and hearing about death affected us all very closely. These were tragedies happening to people I knew.

"The longer the war went on, the less publicized things like the ATA became. After a while the public didn't pay any attention to us. Sometimes we'd sit in a bar having a drink in our blue uniforms which had wings on them, and somebody might ask what service we were with, but in general the reactions were minimal. There were all sorts of other things going on. People were going through a terrible time, being bombed, losing their dearest, their men away fighting, and I don't think anybody thought much about what anybody else was doing. Every time I made a good landing, especially after a difficult flight, I thought, 'That's one in the eye for Hitler' — every little airplane I delivered was one little cog in the wheel against Hitler. But everybody was doing something. People were digging people out of bomb holes, having a horrible time in icy-cold conditions. People were working in factories in ghastly conditions. Ours was one of the nicest jobs you could have, flying all those wonderful fighter planes. We were never bored." In fact, when the ATA decided to take on new personnel in 1943 and train them to fly from scratch, there were thousands of applicants, particularly from women in the air force auxiliary, the WAAF.

Joan Cowey remembers similar attitudes among and towards women in the anti-aircraft batteries. "Despite the early publicity and general bad reputation of army girls, eventually we were pretty well ignored. Nobody knew what we really did, and nobody cared. There was a war on. We still felt great every time we shot something down. We never thought about the pilot — did he have a family or anything. The army trains you to kill. Our instruments and those guns were practically obsolete anyway, and we didn't have that many hits. When we did, the whole group would be ecstatic. It was knowing that you were doing something, part of something . . . you were helping, anyway. To shoot a plane down was something else. We loved it when we had a busy night, just loved it. The men would put up their hands and cheer and the girls were quieter, but we felt great."

"The people who did make a big deal of it, or were the most surprised, were the Americans and Canadians," recalls Diana. "The bigger the airplane, the bigger the surprise. When you landed at an American base they'd be looking behind you as you got out of the plane, to see who the pilot was. They couldn't believe such a tiny girl could handle such a big, heavy airplane." The American military newspaper *Stars and Stripes* ran articles on the brave women pilots and "gunner girls." One piece on the ack-ack women quoted an American colonel: "I have seen some marvelous things in this country, but nothing like the Ack-Ack girls. How calm they seem to be with shells screaming all around them and knowing that any one of these missiles can bring a flying bomb on top of them."[19]

Joan was dating a Canadian soldier. "I met my future husband at a Naafi club one night. He asked me what I did. He was really surprised; he had no idea that girls were doing anything like working on a gun-site. Canadian girls weren't doing anything like that. If they were in the army they worked in an office. We were driving three-ton trucks, carrying one-hundred-pound packs on our backs, working outside all night next to the guns with the constant din of firing in our ears. Canadian girls never did anything like that."

"But some did come over and fly with us," recalls Diana. "There were three Canadian women and twenty-five American women at one point. Then the famous American pilot Jacqueline Cochran organized a similar group in the United States which ferried planes for the American air force. They even flew tow-targets. They modelled their organization on ours but were an auxiliary to the air force, not civilian. They were called the WASPs."[20]

Both Diana and Joan remember feeling tired as the spring of 1944 approached. Joan: "It just seemed as though it would last forever. The novelty had worn off. The spirit of the early years was replaced by a more deep-rooted anger."

Buildup for the allied invasion of Normandy in early 1944 revived spirits somewhat. Diana noticed she was given a lot of ferry trips to the south coast, and she saw camouflaged troops and tanks practising manœuvres in the woods surrounding her base. Joan was sent to a site on the east coast, near Lowersmith, Yarmouth, where she lived under canvas for a few months with forty AA women from different units. "We camped in a huge forest. There were French, Belgian, Polish, American and Canadian units camping there too. We each had our own tents and invited each other back and forth for dinners and parties. We knew the French troops were going over, because a few more kept disappearing each day, so we knew they had been parachuted in. We didn't know where our unit was headed for. We were confined to camp and couldn't go out at all. We were there for some time, and everyone kept saying that Hitler had this new weapon, the V-2, so we'd have to invade. It was all very vague. You are never told anything in the army. We weren't allowed to contact anybody, not even our families.

"The next thing we knew we were on a train. The windows were all blackened out. Nobody knew where we were going — it seemed we must have been on it two days. We arrived at the mouth of the Thames. Some of us got on landing barges; some got on old ships. I was on an old cattle ship. It took us three days to get to Ostend, Belgium, because of the minesweepers going slowly ahead of us. We had cans of Spam and were God-awful seasick. We couldn't even open the Spam, because they had forgotten to give us the keys to open the cans with, so we only ate hard-tack biscuits. That was all we ate for three days. There wasn't a cup of tea to be seen."

Joan had been selected to go to Belgium, along with forty other ATS women, on the basis of good training-camp reports and body strength, and because she had volunteered for an overseas assignment. There were small numbers of other ATS women posted to gun-sites in southern France and the Middle East immediately after D-Day. This decision created more arguments in British Parliament, but the concern was more about the dangers to women on the "wilder shores of love" than about the dangers to them near the front.[21] The reality was that life on the gun-sites was difficult, conditions harsh and the work often dangerous.

"When we arrived they put us up in an old hostel. It was good enough to sleep on the floor for the night. The next morning the trucks hadn't arrived so we began to march to Antwerp, which was quite a way. We had everything we owned on our backs and were just doubled over. As we were marching along there were groups of soldiers coming towards the boats to go back to England. They said, 'Oh, the war must be over, here come the women. It's the end.' When the trucks finally arrived from the AA camp the drivers were upset to see us. Antwerp was really getting it bad. They said they didn't know who was responsible for sending us, but they felt bad about women being sent into it. They told us that they hadn't slept for nights, and they said they didn't want to be responsible for what might happen. It was a very noisy first night at the camp. The Germans were dropping everything, I think.

"One of the first things we had to do was to build a road. Mud was everywhere. They thought the only way to do it was to go down to the docks and get rubble from the bombed-out buildings. The docks had had a really bad time of it. It was dangerous work; our trucks just rocked with the explosions. We had to gather the bricks and bring them back to camp. The girls did this because the men were manning the guns.

"The day after we arrived we had a direct hit. Quite a few girls were wounded. We were unpacking and it came from out of the blue. The Nissen huts just flattened right out. There was an army hospital in Antwerp, but I don't know if that is where the girls were taken. We never heard what happened to them.

"It was so bad there that we had to have a rest leave every so often, just to get some sleep. We couldn't sleep in camp, and we never took our clothes off.

"The girls who got to Antwerp were tough. We were determined; we had to be. We weren't so panicky as the men, I thought. When the V-2 rocket hit our camp the men panicked. I was really surprised. Maybe it was because they'd been through a lot already and were shell-shocked. They rushed around screaming, 'Are you all right?' and we just sort of crawled out and nothing much was said.

"We wanted to be tough so we wouldn't let the men down. I think at this point a lot of the men felt badly about having women stationed with them. You can understand that. There were some veterans of Middle East campaigns, of big battles, and it must have been a real comedown for them to be serving with women. They never said anything, though. It's just what I thought."

Although the British Second Army had liberated Antwerp on 4 September 1944, the Germans held the banks of the Scheldt estuary for the next three months. Antwerp X was the code name for an Allied anti-aircraft defence of Antwerp against V-1 and V-2 rockets. Altogether there were eighteen thousand troops and five hundred AA guns involved in Antwerp X.

The Germans were making their last stand nearby, just across the border with Holland, when Joan's unit arrived in Belgium. Hitler had ordered his generals to hold the Siegfried line and the remaining German position in Belgium and Holland, in hopes that his armies could regroup and launch an infantry offensive through the Ardennes, to Antwerp and on towards the sea. Fighting was occurring only twenty miles from Joan's gun-site in the early months of 1945. Germany's new weapons marked the last German offensive of the war. The first V-1s and V-2s were fired in 1944 from Holland, primarily at two targets — London and Antwerp. The V-1 was a flying bomb powered by a jet engine that made a growling noise and earned the nicknames "doodlebug" and "buzzbomb." It flew in a straight line at a range of up to 250 miles, at about 770 miles per hour. Over two thousand of these rockets were dropped on Antwerp. The V-2 was a more deadly weapon, a fifty-foot rocket controlled from the ground. It flew faster than the speed of sound and arrived on target without warning. Antwerp was the principal victim of the new weapons, receiving 1,265 direct hits from late 1944 to March 1945.

"We were a target in Antwerp," says Joan. "I've never seen anything like it. Just devastation. The V-2s were the worst: you didn't hear them. All you heard was a rush of air and then a violent explosion. They went off around the camp all the time. I was young then, but I couldn't stand it today. It was scary. They almost blinded you with a flash of light. Of course, it is nothing compared to what they have today, but it was really bad, and they were dropping everywhere.

"The most frustrating thing was that we couldn't hit them. They were much too fast. We couldn't move the guns around fast enough. People would yell, 'Move, move . . .' and we'd try, but never in time."

An ATS woman stationed with Joan wrote in her diary: "We have had more doodles and rockets these last few days. Antwerp is having a bad time. Poor Sue is really nervous, and Peggy and some of the others are a bit jumpy. I'm not frightened, not like I used to be, but I shall be so relieved when it is over. We aren't allowed into Antwerp. Everywhere is very dead. There is a lot of damage, and nobody is bothering to do anything about it. There are hundreds of little boys. . . . They are all over the camp and around the streets as their school has been bombed. We were the object of great interest on our walk today. People stopped to stare at us. Children followed us down the streets. It made us feel self-conscious. It is difficult to know how many are well disposed towards us. Lots smile, but lots look at us rather queerly."[22]

Joan: "The campsite was just mud and Nissen huts, and orphaned children wandering around looking for something to eat. Some were only about two years old, freezing cold without proper clothes. They didn't have shoes. They used to get into our swill buckets, where we'd scrape our plates after a meal. The men tried to stop it, but we couldn't stop it — they were hungry. I felt guilty taking food back to my hut without giving them something. The men said not to do that because then we wouldn't be able to get rid of them. We did it anyway. We brought them in and let them sit by the fire to get warm. Lots of us began to sew and knit for them.

"Between us all there developed a strong camaraderie. A lot of the men had seen action somewhere and had been wounded, or were too old for active combat, so they were put on the AA sites. They were good to us. I think they worried about us more than anything. When we got there of course there was no water. You had to get it from somewhere else. They would get it for us and heat it up on an old stove for us outside so that we could wash. They were just great to us. They got bothered if somebody got sick, would take over your duty for you. But when the alarm went, that was a different thing. We were all the same when the alarm went, and they expected us to do our jobs and not to let them down.

"They were really great, I miss them all. We were all friends, all friendly, but not that close — because, well, it didn't do to get that close. You would be immediately posted somewhere else if they found out. We had dances and that sort of thing, but if you were caught with a fellow more than twice they'd think something was going on and post you.

"We proved that men and women can be stationed together with no problems. We were comforting to them in a way; they'd enjoy our company. I don't really know where people get this idea that men and women should not be stationed together. It's just fear of sex. I'm sure it was going on, but so what. Anyway, most of the time we were filthy, hadn't taken off our clothes for days and were exhausted. When we were off duty we were too tired to go anywhere or do anything or even think. We just went to bed. Sometimes we'd invite the men into our hut at night, we'd have a fire going and be in our nightclothes, and make some hot chocolate. Nothing ever happened. We talked about things like a letter from home. If one of us had a pack of cigarettes, we'd share it."

Living among a lot of women also was an experience that took some getting used to. "You always hear people who have to be around groups of women saying, 'Oh, I can't stand them.' Yet we were eighteen in our hut and I don't remember one fight. Arguments, sure — we'd argue about silly things, clothes, chores — but not fight, never. Not with the men either. I guess it was like a family. Sisters and brothers argue all the time, but they also look after each other.

"One night two girls went on leave into Antwerp and didn't get back, and all the men who were off duty went looking for them. They were worried because it had been a particularly bad night for bombs, but the girls had just met a couple of soldiers and were late getting back in. We all felt much better when they got back, though."

"We had been there two weeks when a call came down that our troops at Bergen Op Zoom, about twenty miles away, would like to have some company. They'd been on duty for hours and hours. We went by truck, but when we were almost there the Germans broke through the lines and we had to turn round and go back to camp. That's how close the fighting was.

"I don't think people really knew we women were overseas. In Britain women were doing everything in the war. Don't forget that Churchill said that every man and woman should do something. It was just accepted. There were girls on tractors, in the navy and air force. But it wasn't quite as accepted to have girls going over to Europe and the Middle East. That was hard to swallow in England, I think. My parents, when they found out where I'd been sent, thought it was very foolish of Churchill to let the women go.

"Winter continued, and still the fighting wasn't letting up. It was pretty bleak there, very cold — and the mud! There were days when we couldn't get any wood to light a fire. We were in wet clothes a lot, which was awful. For about three weeks when there were heavy attacks, we slept in our clothes every night, if we slept at all. Our nerves got a bit frazzled. There were nights, after three or four days out in the field, when we'd come off duty, sit up in our hut and say, 'Oh, I wish this would end.' Then getting a twelve-hour leave to go into Antwerp and running into more bombing, having to help pull somebody out of the rubble. It seemed never-ending. Things seemed to get worse and worse, and if the Allies did advance, they would be pushed back too. There were strange stories coming out, rumours all the time, so we didn't know what to believe. It was frustrating and slow going. On the roads there were overturned tanks, trucks and little crosses everywhere. It was all around us, and so many people were dying. In the end it just seemed to bounce off me. I was so involved in it that I couldn't think about it. You had to carry on.

"We tried to keep things normal — like looking good, which was a pretty hard thing to do under the circumstances. We had to keep our hair short and wear battle fatigues and leather jackets. There's not much you can do with your figure when you are in those, but we'd get the odd magazines and things like that. My boyfriend by now was also stationed in Belgium. Sometimes we managed to get out together, though it wasn't easy for him to drive into our camp since a lot of the route was bombed and out of zone."

Diana Barnato had fallen in love with Derek Walker, a Wing Commander with the RAF. Late in 1944, when Derek got leave from his post in Brussels, the couple were married. They flew one of their last missions of the war together. "After the Low Countries were freed, they started to consider whether the ATA would allow women to fly to the continent. It was so stupid. They were saying no, women can't go across, there aren't loos for them — they might get caught by the Germans. Or, the living conditions are too bad. Well, the living conditions in England were bloody bad too — some of the places we got stranded at! We were all very upset that we weren't allowed to go when the ATA men could. Eventually that was got around — partly by me.

"My husband wanted an aircraft delivered to Belgium. He was stationed with Air Chief Marshal Moary Cunningham, Chief of the Second Tactical Air Force of the RAF, in Brussels. Derek wanted a photographic, high-flying aircraft for reconnaissance. He didn't have enough pilots, so he asked if I could bring it over for him. I told him the ATA wouldn't allow it, but he pulled some strings, and Moary Cunningham in effect ordered the plane and me with it.

"I was told if anything were to happen not to land on the continent but to get back to England, because of pockets of German resistance which hadn't been mopped up. No one knew which airfields were safe anymore. It was very exciting to fly across that strip of water. It's only twenty-two miles, but to go to the other side after not being there for so many years. . . . As I came over the coast, I saw the bomb holes in the sand.

"I got to Brussels and had four wonderful days with Derek. We made a lot of friends there and it was so good to be together. I was even permitted to fly the reconnaissance plane over German lines one day. Great fun.

"The day I was supposed to leave, there was a very thick fog. Derek rang up my CO and got me permission to stay an extra day. The next day looked equally dismal, but I couldn't afford to stay away any longer and Derek was in a hurry to get the film to England. He suggested we take off and do a circuit to see if the weather was as bad as it seemed. Sometimes it's not so bad once you're up in it. He said, 'You stick in formation with me.' We flew for fifteen minutes, and I was beside him. The weather was dreadful, not my sort of weather at all. He was cruising at RAF revs and boost — slightly faster and higher than the ATA did — so we were going a little faster than I would have cruised at, which makes a difference when you aren't used to it. All I could see down through the mist was the sun now and then shining down on bits of water. In the fog were lumps of muck floating about. I was hoping Derek would turn back when suddenly his plane disappeared into a lump of muck. There I was in the middle of nowhere. I went down very low to try to get an idea of where I was. He had told me not to hang about because they'd shoot at us. I could see it was lovely farmland, but nothing I could identify in a short time, in terrible weather. I flew over the hills and suddenly the coast appeared. I saw the strip of land with all these bomb holes which I thought was the same strip I'd seen coming in. So I set out over the sea at 315 degrees and thought I would hit the coast in seven minutes. After seven minutes, I looked down through the sea fog, but I was still over water. I knew something was wrong. I thought perhaps I was flying up the North Sea, and I would run out of fuel soon, and nobody would ever know. I changed direction, hoping I might see England and could bail out.

"I was flying very low and in a few minutes I came out of this mist and fog and saw a little white line ahead of me. I thought, ah, the White Cliffs of Dover, lovely. But it wasn't. When I got there, it was the sandy shore and a gasometer, enormous, looming up out of the fog. Then I knew I was near Tangmere, an RAF aerodrome.

"When I got to the airstrip, lights were coming up at me, white Very lights and green ones, and they never stopped. I thought they must be bringing in a squadron, so I circled around for a while. All I wanted to do was to land before I ran out of fuel. I circled and circled, and nobody landed. The runway was still clear, so I decided to land before I lost the airfield in that fog. As always happens when you are scared, I made the most beautiful landing. I taxied in toward the watch office, and there outside the door was Derek's Spitfire! He came rushing out of the office, saying 'My God, Diana, how did you get here? With the dreadful weather this is the only aerodrome in the south of England that is still open!'

"The next day Derek rang me up at Hamble and asked me if there had been anything wrong with my plane. I said no, why? Then he told me that someone had started it up the next day and it blew up on the take-off."

On 7 May 1945 Germany surrendered at Reims. After five long years, the full horrors of which would be known only later, it was finally over. In London the battered streets filled with crowds ready to celebrate with abandon. Diana was with her husband and her father: "Before we went to Buckingham Palace to join the crowds, my father came round with a bottle of champagne. He said, 'I've been saving this to celebrate the end of the war.' He stood by the large open window which looked onto the street and the river, and opened the bottle. He sniffed at it and said, 'It's corked' and threw it out the window. It splashed all over the pavement, and we laughed and laughed."

On that damp spring night in Belgium, Joan's unit sat in the Naafi, scraping the mud off their boots and reading magazines. Rumours that the war was ending had been circulating for days, so no one was really expecting it right away. "Then our commander came in and told us that we might hear that night. We stayed up all night, playing cards, and at five in the morning one of the officers came into our hut and said simply, 'It's over. Here's your champagne.' We went outside and drank it and woke up all the villagers. Troops down the road had heard, and they came by in their tanks and we hopped on and drove into Antwerp. All along the road women were throwing flowers at us, and there were thousands of people rushing into the streets. We couldn't move for the people. When we got into the city, it was crazy. People were climbing lamp-posts, dancing. There was a lot of ugliness too. In the back streets people were turning on those who had been friendly to the Germans. The police had a terrible time controlling it.

"A day or two later, it started to hit us, that it was really over. I can remember sitting out in a field and the major said, 'Put up your hand if you would like to do such-and-such.' No hands went up. We didn't know what to do. We were trained for the outdoors, we could peel potatoes. . . . We weren't good for much else. That was the one mistake they made: they trained us for war, but they didn't prepare us for peace. We were all fresh out of school before the army, and now what good were we? Depression sunk in slowly. Suddenly everyone was scared and sad to be breaking up."

There was little formality when it was over. The Air Transport Auxiliary, being a civilian organization, could even dispense with discharges and medals. In a memorial service in September 1946, Lord Beaverbrook thanked the pilots, who, he said, "were soldiers fighting in the struggle just as completely as if they had been engaged on the battlefront," but some ATA pilots felt the end was a sorry letdown without thanks or acknowledgement. Others were simply glad to be going home. Diana: "Everyone was tired, weary. The war was over and there was nothing more to be done. The ATA disbanded and we quietly went away. I didn't miss it. I was so happy that peace had come, so happy to resume my life, to begin my marriage. I couldn't wait."

The life and marriage Diana had patiently awaited for five years fell apart a short time later when her husband, Derek Walker, was killed in a plane crash.

Joan Cowey married her Canadian soldier in Brussels and remained in Belgium for the following year. She was assigned a secretarial post with the army offices in the capital, a job she found depressing and discouraging. "At first we arrived in our battledress, filthy, looking for a bath. Me and my girlfriends used to think then, is anybody ever going to recognize that we are soldiers? We'd seen the Canadian girls in their smart uniforms with make-up on, which we hadn't seen for ages. It seemed we were the forgotten ones, once the war was over, and all these office girls were walking around looking great.

"Then working in the office was a shock. I'd never been in an office. My job was to go over the casualty lists — those missing or reported dead. It was a depressing job, and being in an office after I'd been out in the open all those years was very hard to take. That's when I got married and got out."

One weekend shortly after VE Day, she agreed to meet her husband in the town of Naimagen: "It was then that the full impact of the war really hit me," she recalls. "I had arranged to meet him in the square in the centre of town, in front of the railway station. I stood there waiting for him when a train pulled in, and thousands of refugees from a concentration camp poured out. Some were still in their striped uniforms, now in rags, carrying little bundles. They looked like they had been dragged back from hell. Suddenly some of them saw me and started to crowd around me in a circle, moving in closer and closer, saying things I couldn't understand. I thought, my God, they are going to tear me apart right here in the middle of town. Just then some Canadian soldiers came to my rescue. They spoke their language and found out that because of my uniform they had thought I was a German woman. I was shaking, and it suddenly dawned on me — all the evil. What these people had been through was just becoming known. It was beyond the imagination.

"When it was over, I really missed it. It was terrible. Even going back to England was hard. One brother was still in Burma, my other brother had joined up, and my kid sister, who had been a child when I left, was grown up when I returned. We were complete strangers. It wasn't the same. I missed my family, the army family. That was the hardest thing, really. I think if they ever do it again they should prepare people for that, because the men had trouble too, when they came back. No one understood them. You wanted to tell people about what had been the most important time in your life. Of course they didn't want to hear — everybody had been through their own war, so you just didn't mention it. When I first got to my mother's after Antwerp I couldn't believe the silence. The first loud noise or bang, and I'd be under the table. That has never left me.

"In January 1946 I left the army. I went into a hut, signed here, signed there and I went out. That was it, that was my goodbye. Nobody said anything. Not a soul. We were left with nothing. I went right to the station. I didn't know what to feel. But that is the British way, you know. It wasn't until later that somebody said, 'Didn't you get your medals?' I said, 'No.' I hadn't even thought of them. So I had to write for them. It was only three years ago that I got my discharge papers. I've never received any benefits, or anything like that.

"I arrived in Canada on the *Queen Mary*. When we landed they played "Here Comes the Bride" on the docks, which was a little silly since some of the girls had two or three children. We got on trains. I was on one for four days. It was July and about ninety degrees, and not at all what we had thought Canada would be like. We were tired and petrified. I came here, right to this street. We really weren't too well accepted at first. I think that is because there were so many of us arriving, and unhappy at what we found, and wanting to go back. The expression for all of us was, 'Well, if you don't like it, go back to where you came from.' A lot of older people resented the fact that we'd taken the men from their daughters. My husband had ditched his Canadian girl for me. That hadn't gone over too well with his family, but they were good to me. Other girls just stayed a few days, turned right round and went back. Nobody ever thought to question why. They came out here, their men were not settled, they were camping out in the Legion, they didn't have jobs. Everything was so different.

"I'd been independent for so long — a responsible person in the army. And then I'd be told I had to go to church on Sunday and wear a hat. I had to buy a washing machine. All the things a good wife did. I felt as though I'd been thrown back in the dark ages. It took a long time to adjust. People in Canada didn't want to know about the European war. They would be grumbling that they had to line up for sugar or butter, when my parents had been lining up for everything for years, when the people in Belgium had been eating turnips, when the people had been in those terrible camps. But you couldn't talk about things like that. I don't know why, but they just didn't want to understand."

LE RESEAU
France, 1939-45

In the secret war against the Nazis, women without number played an invaluable part, participating on terms of perfect equality with men.

M.R.D. Foot, *Resistance*

We didn't think in terms of men and women. . . . We were a family. We thought of ourselves as a small group against the whole world.

Brigitte Friang,
Agent, Paris resistance

I had my handbag in one hand and my pistol in the other. I walked up to a farmer and said, "Good evening." I scared the wits out of him! He was waiting for an agent named Rateau, but nobody had told him Rateau was a woman.

Jeanne Bohec,
Saboteur, Brittany

A weak December sun cast a golden light over the ancient Roman village of Saignan, in southern France. I had gone there to meet Brigitte Friang, a resistance agent in World War II, a concentration camp prisoner and later a prolific author and war correspondent for French television. She led me through the cobblestone streets of the tiny village she now calls home. "The mayor and his wife are friends of mine who were in the same *réseau* (network). They encouraged me to leave Paris and buy a house here almost twenty years ago." We passed by a small square in the village centre. "That wall is where the Germans shot our friends." The ancient setting and tranquillity make the declaration seem unreal, out of place. But Brigitte was no longer with me. She was back, I sensed, in the twilight of the occupation, hearing the firing of shots, once again the nineteen-year-old girl she says she no longer really knows.

A few days later I sat in a smoke-filled, book-lined room of a fashionable Parisian apartment, listening to the story of one of Brigitte's fellow operatives in the war, Jeanne Bohec. It struck me that the war had caused Jeanne, once a country girl, and Brigitte, originally a Parisian, somehow to switch roles. Today Brigitte seeks quiet after her many years involved in war; Jeanne, who had operated in the countryside of Brittany, came to Paris, befriended by all the former Resistance fighters who had acquired personal power along with France's liberation.

Brigitte Friang and Jeanne Bohec are among tens of thousands of French women who joined as active participants in the French Resistance. They operated as couriers, spies, nurses, saboteurs and armed fighters. In the tradition of Joan of Arc, women led partisan units into battle, like the famous Nancy Wake and Pearl Witherington, who were agents of Britain's Special Operations Executive (SOE) and dropped behind the lines to work in France. Some were the organizational heads of other networks, like Marie-Madeleine Fourcade, who headed the Alliance network that operated throughout France and directed the actions of over three thousand agents. During the liberation of Paris women fought in the streets with men. When France was liberated General Charles de Gaulle wrote that sixty thousand resisters had been executed; two hundred thousand had been deported, and of those only fifty thousand survived. There are no separate statistics available on women, but ten thousand French women were deported to the women's concentration camp at Ravensbruck, eighty-five percent of them for Resistance activities. At that camp alone, all but five hundred lost their lives to starvation, disease and the gas chambers.

The motivation that drove women to resist was among other things indignation at the shockingly sudden and massive defeat of France's army, along with an abhorrence of Nazism and the desire to liberate themselves from foreign domination. Individual motives varied with each separate personality; there was certainly an adventurous side to the clandestine activities. Perhaps the breakdown of normal male-female roles encouraged women to discover new capacities within themselves. A male director of the SOE, M.R.D. Foot, later wrote in his book *Resistance* that women "had a way of life in which they were not hampered by a brake on resistance activity that hindered a lot of people: a sense of respect for professional standards and for one's own standing in one's profession. The originality, unorthodoxy, and dash without which a successful resister was lost did not come easily to a great many professional men."[1]

Women in France, as in other occupied countries during the war, proved that feminine skills could be particularly useful in underground work. They were good at distracting and deceiving the enemy, they had the ability to act, they could operate alone more easily than men and they had a strong instinct for danger — a "woman's intuition" — that usually proved trustworthy. Finally, the women who resisted possessed the fundamental quality shared by those of the opposite sex with whom they worked — courage.

Women resisters in France participated in armed fighting to a much lesser degree than did the women in Yugoslavia, Greece, occupied Russia or Italy. When the isolated maquis units became unified and began operations in support of the Allied invasion, many women who had spent years in these groups were denied guns. Although there are no statistics on the numbers of women who actually fought, it would seem that as soon as the Maquis became organized as an army called the French Forces of the Interior, traditional reactions to women in combat came into play. Nonetheless some women did fight, defying the orders against it or operating with more flexible men. Women also operated as saboteurs, working with explosives and participating in armed actions. Those women who operated as liaison agents undertook equally dangerous work, and many were discovered and killed.

When the Germans first occupied France an American intelligence agent described French society as pitifully decadent and found that many of the French secretly supported the excesses and totalitarianism of Nazi Germany. After years of the occupation, witnessing those excesses of brutality firsthand, the general public eventually supported the Resistance to at least some degree. But the dangers of operating in occupied France were by no means restricted to detection by the Germans, and it took a sad but necessary distrust of fellow French citizens to survive. It took courage and determination to join the Resistance, skill and luck to survive it. Women proved to have all of these qualities.

Brigitte Friang is a fascinating woman of sixty who despite a host of war-related illnesses looks twenty years younger, with large brown eyes and a wide smile. Barely five feet two inches tall, and weighing under a hundred pounds, her slight figure has been a source of bemusement to those who know of her life, spent for the most part under fire, fighting or reporting from combat zones. In almost comical contrast to her figure is her deep, husky voice and the hearty laugh that erupts often during conversations.

In 1943, when she was nineteen years old, she joined a network of the Bureau Central de Renseignements et d'Action (BCRA), de Gaulle's organization operating in France. She became liaison officer for the network, and like thousands of other women who functioned in this capacity worked in the open, under the eyes of the enemy, and was doomed from the start.

She has written her own account of those days in the first volume of her autobiography, *Regarde-toi qui meurs.* 1, *L'ordre de la nuit.* Today an award-winning author, she still enjoys talking about the rather exceptional events of her life.

Jeanne Bohec met Brigitte during the war and the two women served in the same network, though in different capacities and regions. Jeanne taught the fundamentals of explosives to agents of the maquis and led sabotage operations in her home province, Brittany. She is also under five feet two inches, and for a moment reminds one of a small version of the statue of Joan of Arc in Notre Dame. Her hair is cut in the same blunt style. A devout Christian, she attributes her survival to some greater will. She wrote of her experiences in the resistance in *La plastiqueuse à bicyclette,* published in 1975. She wrote the book "to record my testimony, and because it was easier to direct people to my book than answer all their questions." But she talks vividly about those days with alternating passion and self-deprecating humour. Sitting over several glasses of Scotch, surrounded by books about de Gaulle, I had the sense that this warm, lively and quite remarkable woman enjoys recalling the days and actions that "affected me for the rest of my life."

Dusk was falling as the German troops reached the outskirts of Brest, a major port in Brittany, on 18 June 1940. Jeanne Bohec rushed home from the gunpowder factory where she worked, grabbed a small bag of clothes and headed for the harbour. "I made my way towards the docks. There were throngs of people, civilians and soldiers pushing their way towards the ships. I asked all the boats' crews if they were going to England. Finally, I found the captain of a small tugboat, a refugee from Loire, who offered me passage across the Channel. There must have been a hundred people crowded onto the boat. Dozens of refugees like me, and Polish and French soliders."

It was twenty-four hours after Marshal Pétain had announced the capitulation of France, and only forty days after the Germans had invaded. In the confusion and chaos of a crumbling society, there were those who immediately chose to flee and continue the fight. One was General Charles de Gaulle, who had escaped on the last flight out of Paris that morning. As Jeanne's tugboat pulled out from the docks, de Gaulle addressed the French people over the BBC, asking them to resist and not accept a definitive defeat. "Has all hope gone?" he asked. "No, believe me when I tell you that nothing is lost for France. . . . This war is a world war. I invite all French officers and soldiers who are in Britain or who may find themselves there, with their arms or without, to get in touch with me. Whatever happens, the flame of the French Resistance must not die and will not die."[2]

The overburdened tugboat reached Plymouth in the mist of early morning. Jeanne and the other refugees were immediately assembled for interrogation by British and French authorities. "They asked us why we were there, why we had left France and who we were. They asked us if we wanted to return. I said that I was disgusted with the senile Pétain and ashamed that we had let down our allies. I told them I wanted to join the army to fight against the Germans. They asked me if I spoke English and I said yes, even though I really didn't speak much at all.''

Jeanne told them the basic details of her life. The twenty-one-year-old daughter of a World War I veteran had grown up travelling from city to city in Brittany because her father was a salesman. She was a former chemistry student who had gone to work at the gunpowder factory when the war began, and she had an impressive knowledge of nitrates. Though she did not tell them that she had always been fascinated by war stories, and especially those about women heroines of World War I, it was clear to the authorities that Jeanne was a committed anti-Nazi. Unfortunately, however, they didn't know what to do with her. There was as yet no policy within the British women's auxiliary concerning accepting refugees, and de Gaulle's vision of the Free French Forces did not yet include women. Jeanne was billeted with an English family for whom she served as a glorified babysitter in return for room and board, learned English, and became more and more frustrated, particularly when the Luftwaffe bombs first descended and the Battle of Britain began.

Jeanne was not the only French woman in England who felt the desire to participate in the fight. Inundated by volunteers and realizing the need to operate an effective auxiliary to the fighting forces, de Gaulle finally authorized the formation of a women's corps six months later. On 6 January 1941 Jeanne was among the first French volunteers to join the auxiliary which would number four thousand women during the war years. It was modelled on the Auxiliary Territorial Services (ATS) of the British army, and the French women received their basic training at ATS schools and donned ATS uniforms. Jeanne was among the first five French women to sign up, and she began her training at the ATS school at Bournemouth with one hundred British women. Certain things took getting used to: ''The first thing that shocked me was the medical,'' she laughs. ''We were all checked for lice! I was amazed at how many girls had them. The food was awful — lumpy porridge for breakfast that I could hardly swallow.''

The French volunteers consisted of refugees, like Jeanne, English women married to French men and eventually women from the French colonies. Jeanne says they did their best to improve the hopelessly ugly ATS uniforms with a little French flair, taking them to tailors so they would fit and using the gas kits as handbags. After the completion of basic training they were assimilated into the French forces, serving as "typists, secretaries or nurses," Jeanne scoffs. "These were things that women were supposed to do in those days. They tried to train me as a nurse's aid. I hated it. I had chemistry and scientific training and had worked with gunpowder. I hadn't escaped France and joined up just to fold bandages!" She went to all her officers and complained. "I was tenacious." After innumerable such encounters she was finally given a job that satisfied her.

"I became a chemist in a laboratory where they made explosives." Jeanne was the only woman assigned to work in this top-secret experiment: "We worked with chemicals that could be bought at any pharmacy, and invented explosives. The formulas are still secret, even today. We worked long, long hours, but it was better than being a nurse's aid." The group worked out of facilities in a country home, setting off detonators on the surrounding lawns. Accidents were common, and Jeanne remembers that all the chemists, including herself, were hospitalized for burns at one time or another.

When the British and French began dropping agents into France to work with the Resistance, they sent men to Jeanne to learn the fundamentals of explosives. The British had a special branch of intelligence designed primarily to train operatives to drop behind the lines in occupied countries, report on the state of resistance networks, spy on German operations and, where possible, co-ordinate the separate and disunified pockets of resisters into an organized body, supplied by and in harmony with London. This special intelligence agency became the Special Operations Executive (SOE). De Gaulle, struggling to assert his authority in a foreign country and his independence of Churchill, countered with his own intelligence unit, the Bureau Central de Renseignements et d'Action (BCRA), which did the same thing. The SOE began employing women agents as early as mid-1941, dropping them into many occupied countries including France. The BCRA was not sending any women.

"I got an idea which I never let up on around this time," remembers Jeanne. "Instead of teaching explosives to men in London, wouldn't it make more sense to do it in France? I asked the BCRA if I could go. I said that I wanted to participate more directly, that this was a brilliant idea. But they said, 'No women, no women.' I kept banging my head against the wall, but I am from Brittany and the Bretons are stubborn. So I asked and asked, to the point where they finally said, all right. I was the first woman to be taken by the BCRA and sent to France."

Jeanne began intensive training immediately. At a school in the countryside she was tested physically, with strenuous physical exercises, including running and jumping over ditches and walls. She underwent mental testing, practised memorization and problem solving in specified scenarios and was taught Morse code. Then she was sent to a sabotage school in Scotland, where she was the only female. "They made me sleep in separate quarters and eat with the instructors. I couldn't do physical exercises with the others." Weapons training included pistols and revolvers, and the Sten sub-machine-gun that was the mainstay of the Resistance fighters. "I was clumsy at first, but eventually was as good as the others," she claims.[3] She learned instinctive shooting, which involves a quick reaction rather than careful aiming, and tested this in a room where moving silhouettes represented the enemy. There were courses on explosives, using jelly-plastic detonators and delayed-action devices.

"We had even more original courses," she later wrote in *La plastiqueuse à bicyclette*: "'How to be the perfect burglar.' A cockney taught the course and I had the feeling that he had genuine experience. He taught us how to open locks without keys. We also learned everything on 'silent killing.' It was a mixture of karate, close combat and the use of daggers. We learned how to put someone out or even kill them with bare-fisted blows. This training left me sceptical about my own chances against a determined and strong man, but perhaps someone caught off guard would find it hard to beat me. In any case the apprenticeship gave us confidence in our potential, and a fighting spirit that would be essential to us."[4]

The next phase of training was at a parachute school in Ringway, England. Jeanne had never been in a plane and wished there were another way to get back to France, but being the only woman among seven men forced her to hide her fears. ''They marked sticks where we were supposed to land. We were divided into teams, French and Polish. I jumped first in our team, and of course the others had to follow. One of the Polish soldiers refused to jump, he couldn't do it. The French men did better, I think, because I was there. When there are women participating, men must live up to them, and certainly not be outdone by them,'' she laughs. ''In general they treated me well, and the men seemed to accept me. They were short of people and seemed happy to get anyone, I thought. Anyway these were at English schools, with English instructors, so the Frenchmen could hardly say anything against it. The English had already sent women into France,[5] so the English were fine. The only discrimination came from the men in the BCRA who in the beginning didn't want women involved. I think it is because France is a Latin country, and even today it is difficult to get equal treatment for women. I don't know why they thought that way. Perhaps they thought women should only do auxiliary work, or perhaps they were trying to protect us. Men go to war, women stay home — you know. But the English didn't think that way.''

In France hundreds of women were already taking a very active role within the Resistance: printing and distributing illegal literature, operating radios, housing Allied soldiers on the run and leading them to safety, transporting arms and money, and fighting within the armed partisans, the maquis. Indeed, the ''Alliance,'' one of the largest networks in France, with over three thousand members, was run by a woman, Marie-Madeleine Fourcade, using a large number of women operatives.[6] Fifty-three women agents from the SOE, and ten from the BCRA, would eventually be dropped into France.

The final phase of Jeanne's training was at a security school where she learned how to live and work undetected in France. Since she had been in England many changes had occurred, and it was vital to know about them. She was given information on various French regions where she would be operating. She was checked to make sure that none of her belongings had been made in England and that she dressed and communicated in an innocent way. Her shoes could not be new, and she could not even use Kotex, which were unavailable in France. She was warned against using any English expressions of speech: "I didn't think I would ever do that," she wrote, "but the first time on the Paris metro, I bumped into a German soldier and said, 'Sorry.' Luckily he didn't understand me."

The France she was returning to in late 1943 was vastly different from the one she had left. The Vichy government and German occupation had imposed strict laws, regulations and curfews. Permits were needed to travel anywhere in the country, and spot-checks for identity papers were made often. Paranoia and suspicions were harboured by the various factions into which the French themselves were divided — those who collaborated with the Nazis, those who passively agreed with fascism, those who disagreed and those who resisted.

The first to go underground were compelled to do so for personal safety. The Jews were the first to become organized, their numbers swelled by those who had escaped Germany and Poland, bringing with them accounts of the Holocaust. A hundred thousand French Jews were stripped of their citizenship and herded into French concentration camps. Their fate was perhaps kinder than that facing the Jews of Eastern Europe, but only because rampant anti-Semitism in the Vichy government resulted in Pétain's request that Hitler allow him to solve France's "Jewish problem" internally.

Others early to resist were the French army officers and soldiers w' ၁ had neither fled France nor joined Vichy, but instead gone underground. The Communist Party, which had been the only major political party to vote against capitulation, formed a large resistance network known as the Red Orchestra. As time went on Nazi policies alienated further segments of the population. The forced-labour drafts that deported hundreds of thousands of men fuelled the Resistance when thousands simply disappeared and joined the maquis in the forests. The Germans, who above all prided themselves upon discipline and orderliness, dealt badly with internal dissent. Their fear of terrorism led to a vicious brutality rather than systematic, workable policies. It has been claimed that Nazi stupidity in dealing with the Resistance saved the lives of many agents.

From the beginning there were those who were compelled to resist for ideological reasons, intellectuals who felt an abhorrence of Nazism. In the early days, before the SOE, BCRA and later the American OSS sent in agents to unify the diverse networks, the academics were the first to visibly protest. The first university demonstration occurred on 11 November 1940 in Paris. Ostensibly celebrating Memorial Day, thousands of students carrying *Vive de Gaulle* banners took to the streets. German troops stormed the crowds and opened fire. No accurate statistics exist on the number of people killed, but over a hundred were arrested and taken to Gestapo headquarters on rue des Saussaies.

Elizabeth Brigitte Friang was only sixteen in 1940 when she began participating in student protests in Paris. The daughter of a middle-class and liberal businessman, she was encouraged by her parents to think and speak freely and to aim for ambitious academic and career goals. She was expelled from her lycée when she began writing BBC messages from de Gaulle on the blackboard during lunch breaks. Her final offence was engraving the Lorraine Cross and a big *V* for *Victory* on a school window. Years later, when she was invited by the same school to speak about her "heroic activities" in the war, she remembers "being happy to decline."

Elizabeth was the only daughter in the Friang household. Her older brother was working in the French administration. Her younger brother joined the student protest with her. "I was upset by the German occupation," she recalls. "I was very young and I wanted to do something, anything, against them. I began university but spent most of my time discussing ways of resisting. We began a very stupid activity at this time. It was very dangerous. We stole weapons from the German soldiers on the metro. We would get up close to them in a crowded train and in the shoving slip their revolvers out of their holsters. It wasn't that hard to do. The city was so crowded then. In the metro people were jammed together. Sometimes we would go into restaurants and steal their weapons from the cloakroom, where they used to leave them in their coats in the early days. We also began to slash German propaganda posters, which were all over the city. It was stupid because the result was not worth the risk we took. We were just trying to find some way of contributing to the Resistance."

Her friends at the university knew that she wanted to become more involved. In September 1943 a BCRA agent arrived in Paris to replace the commander of a Resistance network in the north west part of France who had been recalled to England. Code-named Galilee, Jean-François Clouët des Perushes was sent by General de Gaulle to expand and co-ordinate the network. He was a pilot officer in the French Air Force who had escaped through the Pyrenees mountains in the summer of 1941 and joined de Gaulle in London. Clouët heard about Elizabeth from a mutual friend in England and contacted her when he arrived in France. She quit university, became his assistant, an agent of the BCRA, and adopted her middle name, Brigitte, as her network name. Although she never liked the name Brigitte as well as Elizabeth, she has "been stuck with it ever since." Her code-name in the *réseau* was Galilee-Two.

Brigitte continued to live with her parents in their Paris apartment but did not tell them about her activities. For some time they believed she was still in university despite the strange hours she was keeping and the constant state of exhaustion she was in. She learned to operate radio equipment, which she kept under her bed, and spent most of her nights coding and decoding messages from England. She helped Galilee recruit agents and organize drops of arms and men in the countryside. She was Clouët's messenger and as such knew more than any other member of the network.

"By November [1943] I had decided that it was too dangerous to live at home. There was a curfew in Paris and you couldn't be in the streets between midnight and five o'clock in the morning. If you returned home anytime after ten at night you had to tell the concierge your name. I always came in just moments before curfew, and so every night the concierge heard, "Mademoiselle Friang, mademoiselle Friang," which was dangerous. It was suspicious because I was a young girl and in those days in Paris young girls did not go out every night. Then

I had to go to the countryside one weekend, to arrange a parachute drop. My father called me into his study. He said, 'There is something wrong here. You lead a very strange life. You tell us that you are working at the university until midnight every night. It's not true, I can't believe that. Just tell me, is it personal or political? That is all I want to know.' I told him, 'Political.' He simply said, 'Good.' I think he was more afraid of something personal! And he said, 'Go get packed and I'll send you your mother.'

"My mother came into the bedroom and she asked me if I had packed some warm pajamas and handkerchiefs. At this time it was freezing in France and there was very little coal. She asked me that, but she had tears in her eyes.

"I lived underground. I stayed everywhere — in hotels, with friends. We extended our network through contacts; people led me to people they knew were trustworthy who wanted to fight the Nazis." Brigitte knew how to shoot already but got extra lessons from Clouët, aiming at her own reflection in a mirror. She later wrote, "By the end I could take a gun apart blindfolded. I also learned to fight, where to hit a man with my elbows to knock him out. Eventually I could even throw a grenade a fair distance."[7]

"One of our key jobs was to select locations for the drops and landings. To do that we used Michelin maps. We looked for areas in the countryside, farmers' meadows, land that was far from the road, far from the electrical lines. (A lot of farmers let us use their meadows and agreed to hid caches of arms.) I would take the co-ordinates of the proposed dropping zones (DZs) and landing zones (LZs) and the number of the Michelin map, and I would code this information, cipher the co-ordinates and send this by radio transmitter to London. In London they would check to see whether these had already been selected by another network because there weren't that many ideal zones, sufficiently hidden. If it was clear, they would send back their agreement by radio. Each DZ and LZ got a name, like dog, cat or tree, and a couple of code-phrases. Before each full moon we would send a proposal of a dozen DZs and a couple of LZs. We depended on the light of the moon for the safe landings and drops. If London agreed to send people or arms to a particular one, they would read the first code-phrase over the BBC after the news at 1:30 P.M. and again at 9 P.M. If we heard it twice, we knew the drop was on for that night.

"It wasn't actually my job to be there at the drops, but I liked to go. Our work was so hard on our nerves, always being watched or working right under the nose of the Gestapo, that these events were a relief, a break, a gift for us. To see the parachutes coming down was quite a morale boost. We could see the results of our work in a tangible form." She later wrote, "To us the planes represented London, the mother, the refuge, the umbilical cord, and so we felt that we — the soldiers of the night — were not abandoned."[8]

Brigitte was not there when Jeanne Bohec made her drop in early January 1944. Instead, Clouët waited at the DZ for an agent trained in sabotage and explosives, known only as "Rateau."

When Jeanne received final confirmation of her mission from the BCRA headquarters in London she was thrilled. She was given instructions to operate in the M3 region, her home province of Brittany. She received false identification papers and learned her code-name, Rateau, which means "rake" in English. (All the saboteurs had code-names taken from garden tools.) Then she was given an item that upset her — a cyanide capsule. "Its purpose was to bring death to an arrested agent who feared he would talk if tortured. It was a big, round lozenge covered in rubber. An agent could keep it in his mouth for a long time and even swallow it without danger. All he would have to do to use it was to bite on it and gulp it down. I decided immediately that I would never use it. I am a Christian, and I would put my faith in God to keep me quiet."[9]

Jeanne received a pistol and money to deliver to Galilee. "But the pistol was absolutely tiny," she laughs. "I suppose they thought because I was so small, I only needed a very small gun!" It took three attempts before she landed in France. The first night she boarded the Lysander and sat on the package she was to deliver, as the plane had no seats. Once over France and nearing the drop zone the pilot became worried. He circled a couple of times, but there were no lights from the field co-ordinate he had been given, and he turned the plane back towards England. Later they learned that there had been arrests in the area and no reception group dared meet her.

The second attempt was called off because of bad weather. Finally, for the third time she boarded the bomber that was to take her home. "I was scared, but less than you might expect. There were so many other things to think of that you didn't have time to be afraid, really. I had to jump out of a hole in the plane, not the door, and I only had one parachute, no back-up. The whole trip I kept checking to make sure my parachute was well tied, secure.

"I jumped and managed to land in the wrong field. You've heard of people who land, and the first thing they do is kiss the ground? The first thing I did was to pee," she laughs. "Then I looked around. I saw the lights on about three fields over. I took off my parachute and left it there. I was dressed in civilian clothes and had a handbag which was hanging around my neck. I took out my pistol. With my handbag in one hand and the pistol in the other I started to walk in the direction of the lights. I walked up to a farmer who was pulling out a parachute. I didn't know what to say to him, so I said, 'Good evening.' I scared the wits out of him! They were waiting for this person, Rateau, but nobody had told them Rateau was a woman. So there I stood with my pistol, and he thought I was a Gestapo woman. I told him, 'I've just arrived, I'm Rateau,' and he was very relieved.

"I asked, 'Where is Galilee, the chief?' He was a very tall man, and he saw me walking toward him with the farmer. I was just a very small silhouette. He said, 'What's this, they are sending us children now?' I assured him I was not a child but a woman.

"They took me to a local farmhouse to spend the night. At dawn we left for Paris."

Brigitte later wrote about Jeanne, "London sent in to the maquis in Brittany an instructor-saboteur code-named Rateau. No one knew until she arrived that it was to be a woman, until her funny little voice came from the bushes where she landed. No one could suspect, when they saw her, how much her advice and teaching would mean. Nor did they suspect that when their mission was to jump a car or blow up a train, she would do it with them."[10]

Jeanne and Brigitte met the next day in Paris, along with Clouët and another agent, Patricia. They met at the Concorde metro, and Jeanne remembers that they didn't seem surprised to see that she was a woman. "Being women they knew very well that for the special work that awaited me, a woman was equal to a man," she later wrote.[11] For lunch they went to a restaurant that was crowded and full of Germans. Jeanne admitted she was nervous, and Brigitte told her not to worry, that they always did this. But Jeanne says she was "shocked when they started coding an urgent telegram right there at the table."

Jeanne spent a few days in Paris, mostly holed up in an apartment, too nervous to sightsee in the city she had never before visited. "I was eager to return to the familiar ground in Brittany and begin work."

Her first stop in Brittany was her parents' home in Rennes. "And do you know what my father said when he saw me? He said, 'Well, now you are going to stay home with us.'" She laughs. "I explained to them the best I could and said, 'After all, there is a war on, as they say in Britain.' They were scared for me but understanding. I took my bicycle from their house, and that became my transportation. From then on I began my life underground.

"I was stationed throughout Brittany, but mainly in the south. I stayed in villages with people who supported the maquis, or with Resistance members themselves. I was a Breton, so there was no problem passing as a villager. When the Germans stopped me and asked to see my papers, I would tell them stories like, 'I'm going to see my grandmother who is very ill.' I really had no relations with anyone who was not involved in the Resistance in some way.

"I taught my first student, a young man, everything he needed to know about making explosives in about two hours. In the following weeks I instructed about ten other men. I even taught a priest sabotage techniques. He was later caught and horribly tortured. The Gestapo tore out his eyes and then shot him.

"I moved around a lot, for the safety of the program. In Quimper, I taught in a college and my students were the professors! All kinds of people wanted to resist by this time."

Brigitte was by now acclimatized to clandestine work, but she was finding her nerves constantly on edge. "We were frightened a lot. When we were sitting in a restaurant, if someone just casually looked at us, we thought it was the Gestapo. We were always paranoid — in the metro, everywhere. If somebody was following me, just by chance, in the street, I immediately thought it was the Gestapo. It was very hard on our nerves.

"But the stress didn't make us turn on each other or argue. We were a small group of people who liked each other very much. It felt as though we were a small group standing alone against the world. We admired each other's bravery. We were so fond of each other, it was just like a family. We did everything we could to protect one another.

"But you know, sometimes I said to myself, 'I want to be arrested, because I can't stand it anymore.' When I was in Paris, each day I spent going from one rendezvous to another, seeing people who were coming and going to England. And each time, all day long, we were afraid the agent we were about to meet had been caught and had talked under torture and told about this rendezvous. So if someone was even two minutes late, your heart began to race.

"To compensate we tried to have fun. We were all young and very close. We tried to keep a great *joie de vivre* and our humour intact."

Brigitte's relationship with Clouët was also very close, strengthened by the esteem they had for each other and the long hours they spent together. She wrote: "After a while a sentiment developed. My relationship with [him] was the essential element. It is not possible to spend sixteen hours a day with someone, in an atmosphere of survival, and feel nothing. It is love or hate. These were powerful feelings."[12]

"We depended on each other with our lives," she says. "An example of this trust was demonstrated when Clouët received a summons from the Vichy government. He was in a dilemma. He didn't know if they had discovered who he really was or if it was just a routine thing. He decided he must go to avoid arousing their suspicion, which would be dangerous for his father, brother and sister. He said to me, 'There is a fifty-fifty chance that they know about me. If I come out of the building escorted by two men, take your gun and kill them.' I waited outside the building with a Colt .45 in my pocket. It was very heavy. I was not a bad shot, but Clouët had never seen me shoot. When you think of the confidence he had in me, a nineteen-year-old girl, it is amazing. Luckily, the summons was just routine and I didn't have to shoot anybody.

"At this time men needed women," she reflects. "We didn't think in terms of men and women. Before the war we thought like that, and again after the war. But during it we were one team. The men put their trust in us. Men listened to me, even obeyed my orders. We were a family. We thought of ourselves as a small group against the whole world."

During 1943 approximately fifty thousand French people were arrested on suspicion of working in the Resistance. Thousands were killed. Militarily the year ended on a positive note for the Allies. Advances were made on the Russian and Italian fronts, and Tito's partisans secured large German strongholds in Yugoslavia. De Gaulle toasted the New Year in 1944 by declaring that it would be the year in which France was liberated. But the opening months of 1944 were bleak within France and in the Resistance. Thousands of agents were caught and executed. In the south of France, particularly, isolated maquis groups rejected de Gaulle's orders to lie low and wait for the Allied invasion, instead striking out against isolated German units and conducting acts of sabotage. The German commander in southern France declared that the area was in chaos, and the Gestapo and SS tightened their reign of terror.

Two of the most important agents in Galilee's network were arrested in February and taken to Gestapo headquarters in Rennes. One, Pierre Brosselette, committed suicide by jumping out a fifth-storey window. No one in the network knew this. A famous British agent, RAF Wing Commander Forest Yeo-Thomas, alias "Shelley" or the "White Rabbit," had come back to France to report to Churchill on the condition of the French Resistance before D-Day and had decided to try to rescue Brosselette. He worked with Galilee, and in particular his "go-between, a young girl named Brigitte."[13] Their plan was to disguise two Resistance members as German guards who would drive a Gestapo-favoured Citroën to the prison and say that they had orders for Brosselette to be transferred to Fresnes Prison in Paris. The date set for the rescue was 21 March 1944.

Brigitte remembers Shelley fondly. She, like many others in the network, greatly admired the agent who worked closely with de Gaulle and who has been credited with convincing Churchill to send more aid to the French Resistance. As a gesture of her friendship Brigitte gave Shelley a small puppy shortly before their final rendezvous.

Brigitte was to rendezvous with Shelley at the Passy metro station at 11 A.M. on 1 March to confirm last-minute details for the rescue operation. "He asked me if I could go first to meet someone for him and deliver a message. I said, 'Of course' and arranged this meeting with a young man named Guy for 9:30 A.M. at the Alma station. Then I had to make my way to another rendezvous at 10 o'clock at the Trocadero aquarium before finally meeting Shelley at 11.

"But it didn't feel right. I had always had a strong dislike for Trocadero, stemming from an incident that occurred there when I was a child. I had gone there with my governess and had a problem with a man. I told Clouët about my feelings the evening before, when we were having dinner. He said, 'Brigitte, if you don't like the place, don't go. Send someone else.' They told us in training, and in England, that if you had a feeling, obey it. So I planned to ask a boy I was training to be our chief of liaison. But I stupidly forgot to remind him to go.

"Awakening at 7 A.M. the next morning, I realized my forgetfulness. I thought, 'Well, it's just too bad, I'll have to go myself.' We couldn't afford to miss a rendezvous, because it was just too difficult to contact people if you missed them at the arranged times.

"I went to meet Guy at 9:30 and gave him Shelley's message. We were both early. Guy told me that he had been trying to locate another agent, Georges, but didn't know where he was. Stupidly, I said, 'Oh, I am going to meet him at Trocadero. Come with me.' That was very stupid of me. It was absolutely forbidden to bring an extra agent to a rendezvous. Guy did not tell me that Georges had missed the last two rendezvous with him. If I had known that, I never would have gone.

"We walked there, to kill some time, because it was not recommended to arrive early and hang around and attract attention. When we got there it was just a few moments before 10. I noticed that there was a man waiting in front of the door to the aquarium. It had rained during the night and it was dull and chilly, and I thought 'What is that man doing here on this cold, wet morning?' We began to walk in the garden and we passed two other men who were speaking a foreign language. I said to Guy, 'Oh my God, that is the Gestapo. We are trapped.' Guy said, 'Calm down. Girls are always nervous.' I was furious with him. Then I looked down toward the Seine and I saw two other men walking down the street that cut through the garden, and one of them was making a signal with his hand. I said, 'We'll see who is nervous. You can do what you want but I am going to run for it.' I began to run. I ran to the entrance of the garden and at that moment eight men who were hidden there all jumped out and rushed at me.

"The man in front said, 'Well, Brigitte, we have you.' I said, 'Who are you?' I was trying to act innocent. Trying, anyway . . . hoping. The man at his side laughed and asked for my papers. I had a handbag with a strap, and I began slowly reaching into it when he yelled, 'Hands up!' At the speed of lightning I put my right fist in my left hand, turned left, revolved to the right and gave him a blow with my right elbow to his solar plexus. He fell backwards and I began to run. I rushed into the street, but because it was still early there were no pedestrians. I couldn't hide in the crowds.

"As I ran I was thinking wildly. I knew too much. . . . I couldn't afford to be captured. If I was tortured would I talk? I knew the real name of my chief and those others recruited to go to London. I knew how to decipher telegrams. I knew the places for the drops in the countryside.

"They fired three shots. One hit me in the lower back and came out through my belly. I fell. . . . I was sure I was dead and thought I had gone to heaven. I fell forward, then over onto my back. Somehow I cut my knees. I heard some noises. It seemed strange to hear voices shouting in heaven. I was not unconscious at all. I opened my eyes and I saw the branches of trees against the sky. It seemed strange to see trees in heaven. I heard people yelling around me, giving orders. Someone said, 'Pull her skirt down.' Those moral men were offended that my skirt had come up around my waist. When I heard that, I knew I was not in heaven.

"I was not suffering from the wound, but my knees were really painful because I had cut them in the fall. I tried to raise my back, to look down at my knees, but I just succeeded in raising my head. I saw that my right hand was holding my belly and it was bathed in blood. Then it began to feel painful. I was lying in a pool of blood.

"A German stood over me. 'I'm sorry, mademoiselle, that I used such a big gun, such a big bullet. But it is war, you know.' I answered, 'Thank you. I know.' And nobody believes this but I swear it, I began to laugh. I laughed and laughed at the fact that the Gestapo man was apologizing for the size of his bullet. I wasn't in shock. I just thought it was hilarious.

"One of their cars arrived. A truck drove by, and I can still see the expression on the French truck driver's face. The Gestapo yelled at him to go away, but he was staring at the scene, at me, in horror.

"They put me in the car. There was a driver, another man in the front seat turned back pointing his gun at me and one man on the left side of me in the back seat. I began to feel very ill. I was losing a lot of blood. I told myself that I had to die. The car was going along the Seine, to the east towards the hospital. I thought I must open the door and fall out, and I would die when I hit the pavement. It seemed as though I were doing it very quickly, as I pushed the handle of the door, but it was slow motion. I was very feeble and they were watching me. The man next to me gave me a blow to the jaw. I vomited all over the seat. It made me feel happy to spoil the car with my blood, which was soaking the seat and flowing onto the floor, and to have been sick, too. It is stupid what you think to yourself in these moments.

"I was very feeble. Everything was getting darker . . . darker. I thought I was dying. I wanted to die. We arrived at the hospital; they pulled me out of the car and threw me or let me fall onto the pavement. My head and back hit the pavement, and I recovered full consciousness.

"They took me into the special ward for terrorists, into an emergency room, and put me on the operating table. There were two doctors and three nurses. The nurses began to sponge up the blood I was bathing in. One of the Gestapo men said to me, 'Pull up your dress.' I couldn't and I didn't want to obey. I had a little brooch on my dress, and I tried to tear it away with my teeth. I was hit across the face. The man thought I was trying to take poison.

"The doctors tried to examine me. The Gestapo began asking me questions, hitting me. The German doctors were clearly furious about this way of acting. There was a loud argument in German, which I didn't understand. Finally the doctors put their hands up in disgust and left the room. The nurses put an emergency bandage on me and followed them out. The bullet had gone right through me but missed the vital organs. It was a miracle. A famous surgeon later told me I had a one-in-a-billion chance to survive it. A 9 mm bullet is quite a large one.

"After the doctors and nurses left, the three Gestapo men kept asking me questions. They said, "Where is your boss at this moment?" I looked up and saw a clock on the wall, and I thought, 'I know where he is at this very moment, but you can't get into my brain.' The whole German army could interrogate me, but if I did not want to tell they would learn nothing. And I did not want to tell."

At that exact moment Shelley, the White Rabbit, was arrested by the Gestapo, who had been led to the Passy metro station by Guy. He was taken directly to Gestapo headquarters on rue des Saussaies, where he almost died during subsequent tortures. He was severely beaten and continually submerged in water to the point of drowning. But he did not talk. Later he was transferred to the prison of Fresnes.

Brigitte woke up in a ward reserved for terrorists. She wrote: "It is day. It is night. I am sick. I am so sick. I am thirsty. Ah, for a moment of respite. A minute. A second. I need something to kill the pain. I need some water. They refuse an aspirin to me. Terrorists are not given pain-killers, they yell. The men come back in and surround my bed. Again their stupid questions. They are hitting me. Blows fall about my head and over my body. They aim for my wound. I refuse to cry out. I won't give them the satisfaction. They hit my breasts often during the interrogations. They break my teeth, my jaw. My gums bleed, my teeth crack. I say nothing."[14]

She recalls, "They had my handbag, but it had an unusual clasp which made them think it was trapped, so they didn't open it. They asked me where I was living. For the first time I answered them. I said, 'I live under the bridge at Alma. A man was waiting to write down anything I said, and he actually started to write this, then caught himself. More blows to the face. My God, they were really terrible this time.

"I was beyond fear during these interrogations. They could do anything. During interrogation you could not be frightened because you were 'in action.' People's lives were at stake. You had to protect them — your people, your friends. But I was petrified before the interrogations, when I heard their footsteps outside my cell and heard the key turning in the lock. . . . I was obliged to hide my hands because they were shaking so badly. But when they began to ask me questions I was calm. Some days I would speak a lot, but never the truth. I made up stories. I astonished myself with the stories I made up. I have asked other friends who went through it and they had exactly the same experience. They told astonishing stories. The brain is moving so rapidly with need. . . . It is speeding.

"They threatened that I would get no medical treatment, that I would never walk again. That thought is pretty hard to take when you are only twenty. Then they finally opened my handbag, in which I had my real identification papers. I normally carried fake ones, but on the morning of my arrest I had been forewarned that there would be a large investigation in Paris of all the people carrying cards made in the countryside, where most of the fake ID cards were printed. I had taken my real ones to avoid taking a chance.

"So they knew who I was and where my parents lived. They went to my parents' home, and unfortunately my older brother who worked in the French administration was also home that day. They didn't arrest him or my parents. But they told me that they had arrested the three of them, and said they would be shot if I did not talk. I was in a moral dilemma. I believed they were telling the truth, because normally they did arrest the relatives they found at home. But I chose not to speak.

"Next they told me that four of our agents and Shelley had been arrested on the twentieth and twenty-first, and that they had talked. It was not true. Only Georges, who betrayed me, had talked. They said I would be shot if I didn't talk. But none of us talked. It was easier to choose not to talk because we had all chosen the risk of death."

Clouët and his agents planned to rescue Brigitte while she was still in the hospital. They made arrangements to get doctors' clothes and a black Citroën to use as a get-away car, but the rescue attempt failed and two agents were killed.

Clouët was planning another attempt when he learned that Brigitte was being transferred from the hospital to the prison of Fresnes in the Paris suburbs, where all captured Resistance agents were imprisoned until their execution or deportation.

She was interned at Fresnes for the next seven weeks. During that time she was taken each day for interrogation. "One day I entered the room and the chief interrogator was in very bad shape. He looked at me thoughtfully and said, 'Brigitte, you are luckier than me. Your family is alive, but mine was killed last night in a bombing raid over Berlin.' That is how I learned that my parents and brother had not been shot, and it was a great comfort to me."

She learned that Shelley was also at Fresnes and sent him a message: "Brigitte sends best wishes." Shelley, suspicious of a Gestapo ploy, sent back a message that only Brigitte would understand: "For Brigitte — I have lost my dog." She replied immediately, "Brigitte is very sorry that the great poet has lost his dog, but is pleased to have news of him. She'll try and pass it on to those outside." That was how Shelley learned that Brigitte had also been arrested.[15]

Guy, the young man who had been with Brigitte at Trocadero, had also been arrested. Brigitte: "The valiant Guy who spoke of strong male nerves did not think he wanted to test them under torture. He took the Gestapo from the aquarium right to the Passy metro station where Shelley was waiting."

Through a corrupt Gestapo agent, with money specially dropped in from the BCRA and British War Office, the network paid for the lives of Brigitte, the other agents and the English "White Rabbit." They were sent to concentration camps rather than put before a firing squad.[16] On 10 May 1944 Brigitte left Fresnes, deported to Ravensbruck, a concentration camp for women in northern Germany. The camp held more than 20,000 women of all nationalities at any one time during the war. In total, 130,000 women passed through the gates to face forced labour and death from starvation, disease, the gas chambers and medical experimentation. Before Ravensbruck's sole occupation became extermination, thousands of women were sent on to Dachau or Zwodau's gas chambers after they became too weak or diseased to work.

"It is still too difficult to talk about the camps," she now says. "I wrote about it in my book, because it was easier to write about than talk about. I never talk about it, even after all these years. For two or three years afterwards I dreamt each night that I was back there. Now I dream about it only once or twice a year. When I published my book my friends said, 'Brigitte, how could you write for all the world to read about the experiences you would never talk to us about?' But it was the only way."

The journey to Ravensbruck took five days, the women crammed into a cattle car without food or water. Several women died en route. They arrived at night. As dawn came, the scene revealed itself — row after row of long green barracks surrounded by barbed wire, a watch tower and the SS guards. The gas chamber was farther down the road, disguised as a recuperation clinic for hospital patients.

The French contingent, housed in Block Fifteen, numbered ten thousand women by the end of the war. Eighty-five percent of them had been arrested for Resistance activities and were therefore despised by the SS. Upon arrival Brigitte "discovered the worst, the very worst in people." It was evidenced in the prisoners' faces, their shrunken bodies, shaven heads crawling with lice, covered in sores, sunken eyes revealing only death. "It was a major discovery for a young girl of twenty to find out how evil the human race can be, to see the absolute worst, far worse than anything you saw in combat. People without hope. Complete despair. Starving. Struggling for every breath, for a piece of dried bread or a piece of grey turnip. People would kill each other for that. I saw people licking the ground where some soup had spilled."

The all-woman camp had begun as a prison for common criminals. Its first inmates were German, and some became guards. "They were chosen because of their cruelty and baseness," says Brigitte. "They were the brutal sadists that people who have never before had power can become when they are given it. I don't consider them revealing of women's nature. They had already become debased as prisoners, and despised." Brigitte wrote of her first days at Ravensbruck, "My wound is still running. It won't close. I haven't slept in seven nights. We are all tortured by thirst, worse even than the hunger. I am at the end. I believe so, at least. I do not yet know the limits of human resources."[17]

Preparations for the Allied invasion at Normandy were gaining momentum in London and France. In March the diverse groups of armed resisters in France heeded de Gaulle's request and unified as the FFI, the French Forces of the Interior. Jeanne was teaching the use of explosives to large numbers of new members and also arranging drops of armaments from Britain. Brittany, one of the most densely populated provinces in France, was swarming with Germans, up to 150,000 troops at any one time. The FFI in Brittany were charged with keeping those German units diverted and blocking their route to Normandy once the invasion occurred. All railway lines into Normandy had to be cut in as many places as possible in a simultaneous action code-named "the Green Plan," which was to go into effect on the night of 6-7 May. Jeanne was in charge of cutting the line from Dinan to Questembert.

The day before the Green Plan was to go into effect, the armament drops that were to have included detonators arrived minus the detonators. Jeanne distributed what they had among the other groups and quickly began to make her own detonator. Her contact at the local pharmacy provided her with supplies, but time was running out. Finally, just before midnight on 6

May, she finished her work. The message was read over the BBC — "Reeds must grow, leaves rustle" — and the green light was on. "My stomach was in knots. I was in charge of a group of four of my students, and I kept thinking about the curfew and what we would do if we ran into a German patrol.

> Nearing midnight my team of five, armed with pistols, climbed into a truck. We drove without headlights, taking country back roads. In the silence, the motor sounded incredibly loud. Suddenly we saw a cart emerge, driven by a peasant. What was he doing there? Spying? Black market? He was more frightened by the encounter than we were. He drove his cart into the ditch, and we passed by. We stopped the truck a little ways from the station.
>
> We decided to work on a part of the line about fifty metres from the station. The others covered me as I placed the charges in the switchings and attached them. I drove the detonator deeply into the explosives and joined them with a detonating fuse. Finally, I placed the delayed action, which would give us thirty minutes. I took one last look to make sure everything was in order, and then we took off. We hid in some bushes, and waited. . . .
>
> A half hour can seem an eternity. Someone lit a cigarette. Thirty minutes came — no explosion. Thirty-five minutes. . . ."It's failed," someone said. At that very moment there was an incredible explosion that rocked the countryside. Dogs began to bark. That was good enough for us. We rushed back to the truck, and once again took the back roads to Plumelec and our houses. . . . My heart was beating loudly."[18]

The next day Jeanne learned that all the other teams had succeeded as well. "It was the first combined action of the Resistance, and the Germans were enraged," she says. "They increased their searches and inspections."

The first Allied invasion of France occurred in Brittany on 5 June, only one day before the landings in Normandy. Allied Jedburgh teams parachuted into the region in uniform. These were teams of three intelligence operatives, one British, one

French and one American, sent in to help the FFI. Hundreds of inhabitants of Morhiban rushed to the fields where they landed. Farmers, their wives and children gawked at the first Allied soldiers in uniform they had seen in four years. "Someone rushed in and told me that there were parachutists who had just landed 'in uniform!'" says Jeanne, "'They tell us they are from Great Britain, but we are not sure, they might be German spies. Come and see.' So I went down to the field to see them, and they were the real thing all right. Everyone was very excited."

The following day Jeanne was en route to another mission when she learned that the Allies had landed on the beaches of Normandy. "I was in a train compartment, so I couldn't show any emotion," she recalls. "But it was hard to hide the excitement and joy." Brigitte heard the news in Block Fifteen, Ravensbruck. "The women cried and embraced each other. They spoke of being back in France within a month. But conditions grew steadily worse in the camp."[19]

After D-Day Jeanne worked directly with the FFI, living in their forest camp outside the town of Saint Marcel in the Morhiban region. The Jedburgh teams were hastily training new FFI recruits in the rudiments of weaponry while close to twenty-five hundred men and women worked in the camp, building huts for headquarters and a hospital. Parachute canopies functioned as tents, and in the centre the French flag flew. Large shipments of jeeps and armaments recently dropped were scattered through the base, and the fighters donned armbands in lieu of uniforms to give a semblance of military life. Then, on 18 June, a date that Jeanne found personally significant because it marked the fourth anniversary of her flight from France, a German patrol discovered the encampment. The FFI prepared for action. In a last-minute sermon under a parachute canopy, a priest asked the fighters to pray for "retaliation rather than vengeance" and said in closing, "Let us prepare for combat!"[20]

"They attacked at dawn," says Jeanne, who was stationed in a local farmhouse with her unit. "We were awakened by gunfire. First they sent a company, then a regiment and finally a division to attack us. Fighting spread quickly through the forest, farms, wheat fields and country roads." Jeanne worked at coding an urgent telegram to London, telling them what was happening and asking for instructions. Then she approached the head of one of the fighting units and asked for a machine-gun. "I wanted to be fighting myself. I can't remember who I talked to," she says, "but he sent me away, saying that he didn't want women fighting when there were so many men available. I was furious. I knew how to fight better than most of the men, who had only been recently and haphazardly trained. I believe women should be allowed to fight: it is our right to defend ourselves and our country just as it is men's right. I argued but eventually left. I could have persisted more, maybe I was too afraid, I don't know. Instead I began supplying the fighters in the closest combat area with grenades, and went back and forth delivering ammunition and messages. I was given a message to deliver to another unit, informing them when and where new parachutists were to arrive.

"Another agent and I set out on our bicycles along the main road. I stashed the message in my bra. Suddenly some German soldiers on horses approached and stopped us. I was praying they wouldn't search me. Then I remembered I had a map in my bag which had a lot of fingerprints on it around the Saint Marcel camp area that could be detected if they looked closely. I tried to stay calm. I caught my breath. I smiled at them. They asked me why I had a map. I told them the story about visiting my grandmother. I said I didn't know the area well. My heart was booming, I thought I would suffocate. I looked calm. Finally they told me to move on. I got on my bicycle and pedalled down the road. When I reached my friend I had to stop and sit on the grass because my knees were shaking so badly. It was fifteen minutes before I could move."

The fighting continued with only brief pauses throughout the day. At two in the afternoon the Germans launched another attack with reinforcements. The maquis held them off for another thirty minutes, until the first RAF planes appeared overhead and began to bomb and strafe German positions, including their watchtowers in the town clocktower and a local windmill. Fires spread through the forest and town.

At eight o'clock in the evening the maquis began to evacuate.

German troops almost surrounded the encampment but miscalculated its size and left an escape route open to the west. The Saint Marcel fighters carried the approximately sixty wounded on their backs to safety in various farmhouses in the area. The others carried what arms they could and finally blew up the remaining three thousand tons of ammunition, then dispersed in small groups throughout the region. Five hundred German soldiers were killed in the battle, and only thirty maquis. But civilians from the villages nearby were also killed and wounded in the crossfire and suffered severely in the following reprisals. Jeanne later wrote that she thought the whole operation had been a mistake, that it would have been sounder to have continued using guerrilla tactics than to have gathered such a large force of fighters together, some of whom were very poorly trained. "Even if there weren't too many French losses at the time of the battle, it was not the same at the time of the reprisals that followed. How many of our Resistance friends were killed after? And what of the victims among the civilian population?"

In the days that followed, more than six hundred civilians in a nearby town were massacred in their church. In total over two thousand people, both maquis and civilians, were rounded up and shot in reprisal for the German losses at Saint Marcel.

The day following the battle Jeanne hid with nine others in a farmer's barn, though it was now more difficult to find farmers willing to harbour them. She was expected at a rendezvous that morning but held off her departure. "I don't know why, but I didn't feel like going at all." She delayed for two hours, and when she finally arrived at her destination, a miller approached her, his face ashen. "'Your friends were all arrested by the Germans this morning,' he told me. I was to have been with them." The men were all shot and the women deported. Only one woman survived deportation.

A few days later three other maquis friends of Jeanne were captured at a road block and shot by the Germans. Again, Jeanne had intended to be with them, but at the last moment decided to ride her bike. Distraught, she moved on to Quimper and looked for a place to stay. To stay at a hotel she needed a permit from the police. The German who filled out the appropriate forms seemed sympathetic. He told her there were lots of rooms upstairs, why go to a hotel? "It was absurd, but I didn't hesitate long. Where would I be safer? The enemy would never imagine they could shelter one of their most determined adversaries. It was one trick I enjoyed playing on them."[21]

Caen was liberated on 9 July. Jeanne knew the end was not far off. Finally, on 3 August, the BBC message they had waited for gave the maquis its signal for general insurrection in Brittany. The units took up their posts. The plan was to attack the Germans everywhere and prevent them from escaping.

"I went to get a machine-gun, or at least an American rifle, at the command post. There were plenty of munitions and grenades. The FFI commander didn't mind, but the captain of the Jedburgh team said, 'No, this is not a woman's affair.' I had planned for this for so long. I had even made something of a uniform for myself to wear. I had been given a cap by a French policeman and an insignia from an English sergeant. I didn't want to stay in the background — but they were refusing me a gun. I was very upset, but I stayed beside the men who fought. I rode in their tank and helped fire the bazooka at the retreating Germans.

"It wasn't the men I fought with who felt that way. I was considered a comrade, just the same as them. It was the men in uniform, the professional army. They couldn't accept a woman carrying a gun, and that was it."

Quimper was finally liberated. On the morning of 8 August, the soldiers of the Resistance marched into the centre of town, the flag was raised and the crowds cheered. Jeanne Bohec, alias Rateau, marched among them dressed in her home-made fatigues.

Brigitte had been transferred to Zwodau concentration camp in Czechoslovakia, where she was dying. The day she learned she had tuberculosis, she cried, the only tears she shed during these months. "I thought of my parents. I let go. God have pity. My eyes burned. I was finished."[22] She turned her last resources to the effort of breathing.

She had endured "the constant humiliation, sometimes unbearable. Hunger to the core, fatigue beyond imagination. Classic illnesses and strange diseases . . . lack of sleep which burned eyelids, everything that turned the body into perpetual pain. This vortex, without end, of cries, of blows, of filthy bodies and faces grimacing with hate. The lice that gnawed at that part of the human being one wanted to keep at all costs. The question mark of fate. The filthy, icy, damp misery."[23] But now, as she lay in the infirmary, she knew that if the disease did not kill her she would be transported with all the other "useless mouths" to the gas chambers.

Her angel came in the form of a camp doctor who saved those she could despite personal risk. She moved Brigitte to a cot near the window where she lay delirious, imagining that the trees she saw were the chestnut trees that lined the broad Paris avenue where her parents lived. Sleeping near the window made breathing slightly easier. When the SS came to take the sickest to their death, the doctor told them Brigitte had a contagious disease. The men came no closer. She saved her life over and over again.

On 23 January 1945 Brigitte turned twenty-one. Still near death, she was visited by her two close friends. They had managed against all odds to find a potato, which they topped with a stick, and she described this birthday cake as the "most beautiful gift I ever received."[24]

By the end of March the Soviet armies were advancing into Czechoslovakia. The SS increased transportation of inmates to camps where the mechanics of murder were better able to deal with large numbers. On 16 April Brigitte and her two friends joined one of the last convoys of seventeen hundred prisoners to leave for Dachau.

Along the march rumours were circulating. . . . American troops were only eighteen miles away, Russians only twenty-three miles. But the fear was that the Allies would not arrive in time for the inmates en route to their execution. "The long column marched for three weeks along the 450-mile route, flanked only miles away by the Americans and Russians. So many of us died on those roads. Out of seventeen hundred people, near the end there were not two hundred left alive. My two friends and I had tried to escape several times, and finally we broke away during the night. It was May eighth. The armistice was signed that day, but we didn't know. We didn't know that Hitler had been dead for five days.

"We were in the Russian zone and wanted to move west to find General Patton's troops. All the inhabitants of the Sudetenland were doing the same. The roads were crowded with people, farmers with their carts filled with whatever they could carry.

"As we continued I saw one of the SS guards from our camp sitting on a stone by the side of the road. He saw me and ran. He hid among the Sudetes. They helped him get into their carts. I had a knife in my pocket. I wanted to kill him. My two other inmates and I, along with some French POWs we had met, stopped all the carts and searched for him, but we didn't find him. If I had found him I would have killed him. I am absolutely sure of that."

Brigitte and her friends found the American troops and through international bureaucracy were turned over to French officials, who were in charge of the deportees. Of the 10,000 French women interned at Ravensbruck, only 300 remained when the Swedish Red Cross arrived there, and a few more like Brigitte survived transfers to other camps. In total, of the 130,000 women who passed through those particular gates to hell, 65,000 died or were gassed. Thousands more died in the years following as a result of their internment — years of near starvation, disease and the mangling of their bodies at the hands of butcher doctors.

The journey home began. After waiting for several weeks Brigitte was finally sent back to Paris by train. She arrived there in the middle of the night and was taken to a French army depot where she was questioned by a French officer about her Resistance activities. She waited in the cold, lonely offices until morning. She wrote:

> I am not allowed to return home alone. A French scout must accompany me. The relief has not arrived. I have to wait. Finally, to make myself wait patiently, I sat down at a table where I could get breakfast. But the day cooks had not arrived yet. So they gave me a slice of cold, rare roast beef from

the night before. Red meat is good for a camp prisoner. But at six in the morning, why had I dreamed of scalding coffee, real coffee and warm croissants, or simply good Parisian bread with fresh butter. . . . Dreams. Didn't I know yet that they are stupid.

Finally, she was taken home:

I hadn't included the metro in the dreams of my return. . . . I explained to the nice scout that we'd have to change lines at La Motte-Picquet and then at Michel-Ange-Auteuil. Then I fell asleep. We had to get off, change and then change again. I felt like lying down on the bench and sleeping. To sleep the deepest sleep possible before hearing the "aufstehen bett machen" ["stand up, make your bed"]. A nightmare. . . . I was still in the camp.

Then the familiar station, Pompe. At the top of the stairs I asked my guide to leave me. He hesitated. He had his orders. But I wanted to be alone. I didn't want to share with anyone, not even a shadow, the noise of my feet under the sumptuous chestnut trees. Three hundred metres away, my parents are sleeping. I want to re-enter the world of my childhood, to savour drop by drop the last seconds of my return from hell, those seconds I've been dreaming of for fourteen months."[25]

Jeanne Bohec was assigned back to England after the liberation of Brittany to work at the BCRA headquarters, where she helped in the office until the final liberation of France. She reached Dartmouth the day after General Leclerc's forces entered Paris on 24 August 1944. She returned to France when the BCRA moved its headquarters to Paris in late September 1944 and was not demobilized until August 1945. A short time later she married a Frenchman she had met in the Resistance. On 30 May 1945 she was presented with the Military Cross by General Charles de Gaulle. She later received the Resistance Medal, the Cross of the Chevalier of the Legion of Honour and a diploma from General Eisenhower for brave conduct. Her marriage ended shortly after the birth of her only child on 19 June 1946.

She says, "We had mistaken our identical way of seeing the resistance against the enemy for extreme harmony. Peace proved us wrong." Jeanne began teaching, talking little to friends about her role during the war but dreaming of it almost every night. "Sometimes during my sleep I find myself face to face with the Germans and just about to be arrested. And if my body does not have scars from that time, my soul does, wounded by the memory of all my friends who died around me."[26]

Brigitte returned one month later than most of the deportees. Her father, believing that she was dead, had not spoken for two months. "Returning from hell, I felt so old," she says. "My parents seemed like babies to me. They would never mention the camps, the war. It seemed childish. I felt a thousand years old. We got back to civilization and we were not understood. But we did not want to be understood. We were like people coming from the moon . . . we knew a different reality. I was so different. I was not like my family anymore. I was like my fellow deportees.

"There were many episodes. . . . First I had to report to the army to receive my back pay and benefits. The benefits I should have had — my monthly pay, uniform allowance, laundry, food and so on — of course I had never received. But the man there just gave me my back pay without benefits because, he said, 'You were clothed and fed by the Germans.' My first reaction was to grab him by the throat, but I said to myself, 'It is not his fault. It is the Ministry of the Army's fault. This poor man is just doing his job.' No one understood us. No one wanted to talk about it. They shut their eyes. They forgot.

"Just after coming back I went to the bank. There was a long queue because the government had changed the bank notes. I began to faint. People surrounded me and asked me what the matter was. I said, 'I'm not feeling well, I've just come out of the camps.' They said, 'You must go to the front of the queue, you go first.' Then I heard a man saying loudly, 'Why didn't she stay in her concentration camp!' I fainted."

Brigitte says that once she had regained some weight and physically recovered, people thought she was "completely better." "But the things we had dreamed of, the clean sheets and perfume, did not prevent the nightmares. When Clouët came to visit me at my parents' home when I first returned, he said, 'Forgive me, Brigitte. If I had known about the concentration camps I would never have paid the money to send you there.' I said, 'I will never forgive you.' ''

At some point in the years that followed, she ''became reconciled to the human race'' and ''of course I forgave Clouët.'' But the war had irrevocably altered her life. She retained an obsession with death, and her feeling that only her war companions could understand led her to live her life in war, as a correspondent, where she could analyse war's causes and effects. Perhaps more, she could observe its effect on human behaviour, witnessing the worst and sometimes the best human beings are capable of. After many years as participant in, and witness to war, she still believes it was in the concentration camps that she discovered the worst in human nature. It has remained her most profound discovery.

She told her friend André Malraux that returning from the camps was like Dante's return among the indifferent.[27] No one could comprehend, and few were willing to try. ''I think something was added to my personality . . . a certain relativity. I am always thinking of death — ever since the camps, where it was my companion for fourteen months. We never stopped seeing it. We were united in our attempt to evade it. It meant that there was no direction when we got back, back among the indifferent. Was that reality, or was reality back in the camps?''

VIA RASELLA
Italy, 1939-45

They say that women have babies, and so don't kill. . . . At that time it was clear that each Nazi I killed, each bomb I helped to explode, shortened the length of the war and saved the lives of all women and children.

Marisa Musu,
Gappista, Rome

Carla Capponi's presence among the partisans was very much felt and it could not have been otherwise. . . . Soon in Centocelle was born the legend of this young blonde woman who went out at night to shoot Germans.

Rosario Bentivegna,
Achtung Banditen!

We stood outside in the cold winter air saying goodbye. In the light of late afternoon Rome's gold- and sienna-lit buildings were even more beautiful than usual. Carla Capponi wrapped her layers of wool scarves and sweaters closer before walking off, a tiny woman followed by her enormous black dog.

We were only blocks from where a bomb had exploded on a day like this one forty years before. It was planted by a young woman known as Elena. She had waited until the area was clear of pedestrians, even flirted with two German soldiers before lighting the fuse and walking, unhurried, away. She continued walking without turning to look back while the explosion rocked the street. As I watched Carla Capponi disappear into the crowded piazza on this winter's day, I could almost picture her, Elena, as she had been that day forty years before.

As Germany's ally in the war, Italy had to face disgrace on the battlefield and unrest at home. In 1943, for the first time, the Italian public openly protested Mussolini's leadership and organized strikes (which were illegal) demanding compensation for damages from Allied bombing, and greater bread rations. Women were the first to go out on strike, and several were arrested and threatened with deportation to Germany. In July 1943 Mussolini was expelled from office by Italian King Vittorio Emanuele II, and Italy surrendered to the Allies. But German forces overran two-thirds of the country, and it was then that armed partisans began operating throughout Italy.

Approximately ten percent of the partisans were women. In Italy's long and turbulent struggles against foreign occupiers and its own tyrannical rulers, women have always played a role as revolutionaries. Women guerrillas are said to have fought for Italy's independence in the mid-nineteenth century, and along the northern borders of the country during World War I. Though denied a role in the Italian armed forces, they have always taken some part in the country's conflicts.

The partisans who lived in cities operated underground and took code-names. Carla Capponi (Elena) was then a twenty-one-year-old translator and typist in Rome. She became the most famous woman in the Italian resistance for innumerable armed actions against the Germans. "Rosa," an eighteen-year-old student in 1943, was another member of the groups of armed partisans who operated in Rome. Rosa is Marisa Musu, a respected journalist who still lives in Rome. Carla and Marisa participated in some actions together, including the most famous partisan action in Italy's resistance, an ambush of German soldiers on a narrow street named Via Rasella.

Frantically overworked, Marisa Musu could agree only to a short interview at her office. As my brother (and interpreter) and I entered the room, a short, dark-eyed woman with a youthful face and vital manner stood up and shook our hands. She had a brisk business-like air and seemed a little hurried. Within moments, however, Marisa had us laughing and relaxed. She spoke rapidly, without hesitation. In our short time together she told us, virtually non-stop, about herself and her experiences during the war. She was so very young, only fifteen when she began intelligence gathering for the underground in 1941, that her actions seem all the more remarkable. I looked at a wartime photograph of a sweet, shy-looking girl, her eyes downcast, in a tattered dress. The photo revealed nothing of the determination she had then and exudes now. Before we parted I asked her why she had never written an autobiography. She answered, "Who has time?" laughed, and added, "My life is not over. Maybe I'll think about it when I'm eighty."

We met Carla Capponi at the Association of Italian Partisans (ANPI) offices in Rome. When she arrived, several other women who had been telling me about their experiences as partisans became noticeably excited. They all rushed to kiss her. She looked over at us and smiled. With her dog Marta in tow, clothes dishevelled and hair hastily pulled back, she made a picturesque and endearing impression. Her enormous blue eyes flashed. We sat down, and Carla began her story, gesturing emphatically. The other women listened, entranced though they had heard her story many times before. The hours flew by, and after a while I almost believed I could understand Italian.

Across the street from Palazzo Venezia and the famous balcony from which Dictator Benito Mussolini publicly declared war on the Allies, a young woman posing as a music teacher ushered a continual procession of people into her apartment. In the dimly lit, antique-filled rooms of the stately if impoverished Capponi household, groups of left-wing students met to plot resistance to fascism and the unwanted war.

Mussolini had declared war on the Allies on 10 June 1940. At that time Carla Capponi was seventeen years old, working as a secretary in a chemistry laboratory to help support her widowed mother and younger brothers and sisters. She joined the Communist Party, the most active and organized protest group operating in Rome, and her apartment became the central meeting place for clandestine planning and organization.

Carla knew from her school days many of the people who gathered in her apartment. Raised by liberal parents, she had attended a high school, run by a Catholic Communist, that "produced the most prominent anti-fascists." She had stayed in touch with them and they remained her closest friends. She explains that she, like they, "wanted to change Italy into a more civilized, freer, more democratic nation where there would be greater social justice. We were young, and young people are affected by injustice. They possess qualities such as generosity, and feel sympathy towards their fellow men. That is what drove us to organize against fascism — and later to fight."

Even had she been apolitical, Carla would have attracted attention for her magnetic personality, great sense of humour and personal warmth. Her looks were striking — pale, aristocratic face, blond hair and blue eyes that appeared more Scandinavian than Roman, a flair for exotic clothes and penchant for pipe smoking "to save on tobacco." But her political convictions had become the dominating force in her life after fifteen years of life under fascist suppression by brutal violence, censorship of the press, abolition of labour unions and loss of civil rights.

"As opposed to Poland or Czechoslovakia, which were occupied by the invading enemy, in our case liberation was more an ideological thing. Our first invaders were the fascists, so our initial rebellion was directed towards a political system which had taken away our freedom. Apart from this, what really motivated us was aversion towards the war, the Nazis and their violent power."

Another of the young people who attended the Communist meetings was a fifteen-year-old student, Marisa Musu. She shared with Carla an upbringing by liberal parents who openly discussed their political beliefs with their only child. Like Carla, Marisa joined the Communist Party after the war began because she believed in active protest: "My mother and father actively supported left-wing parties," she says. "At the beginning of the war I decided to become involved in the resistance and had to choose which anti-fascist movement to support. I chose a different political movement than my parents. I don't know if this was due to the natural opposition young people have towards adults, because I was very close to my parents, but by 1941 I was involved in clandestine activities with the Communist Party. It wasn't an ideological or theoretical choice, but a rather childish one, due to the fact that at the time it was only the Communists who were doing anything about fascism." Marisa, then a tiny girl with short dark hair and large dark eyes, says today that she has always been a pragmatic person. "I really thought that my parents were just wasting time discussing what to do. I believed that what counted was action, not words."

The two young women worked as liaison agents for the Communist Party. "We formed a female section of the organization with the help of several women who had just been released from prison," says Carla. The women helped to co-ordinate activities with other protest groups including socialists, the Action Party, Christian Democrats, liberals and labour groups. They printed and distributed subversive literature, helped organize public demonstrations and strikes; and aroused the public, which had never fully understood the need for war despite Mussolini's anti-French/English propaganda, and which for the first time in fifteen years was questioning Mussolini's motives.

In 1940 Mussolini had deluded himself that the country was prepared to launch a "lightning war" against the weak, pacifist French and English. He believed, or so he said, that twelve million men armed with superior weapons stood ready to fight for Italy. The truth was that fewer than one million poorly equipped and mismanaged soldiers had been recruited to army regiments that did not even possess tanks. Skilled at exercising authoritarian power by manipulating and misrepresenting facts, Mussolini now envisioned spreading his power throughout the Mediterranean and North Africa simply by force of will. He told the Italian people that war was "glorious, heroic and profitable" and that fascism had "perfected the art of total war."

Ironically it was the war that he was convinced would profit Italy and expand his personal power that finally revealed the fallacy of his strength and the extent of his deception. Italian armies, ill-prepared to fight a war on one front, were ordered to attack France, Greece and then North Africa in the course of six months. Even as troops suffered humiliating setbacks in Greece and North Africa, Mussolini ordered thousands of soldiers demobilized during the first winter to save his coffers. Government press releases announced Italian victories "unprecedented in history"; the first troops arrived home and told a different story. Italians, confused and suffering economic deprivation, mismanaged to the extent that at one point they had the smallest bread rations in Europe, were shaken to learn about their military defeats. Carla remembers, "After boasting he had nine million bayonets and a strong and powerful army — all this sense of grandeur, of fatherland and empire immediately crumbled on the first encounter with the Greek army. Our soldiers came back defeated, and if the Germans hadn't invaded Yugoslavia and Greece, we probably would have lost completely. They came back and told us how poorly equipped they were, from unsuitable shoes to inappropriate clothing. This created disillusionment and deep resentment in the hearts of every mother and every Italian. The resentment continued and increased when we heard more details from the soldiers who had fought abroad and seen the cruelty of the Nazis. They described how German soldiers surrounded the villages in Yugoslavia and captured all the young men, blinded them, filled baskets with their eyes, left them in the middle of town and then burned down their houses. We had not yet learned of the concentration camps, but already felt a strong revulsion."

On 11 December 1941 Italy and Germany declared war on the United States. Mussolini scorned the U.S. as being militarily weak and incompetent, contaminated by Jews and blacks. Thousands of Italians hoping for an announcement of peace arrived at the Palazzo Venezia to hear instead Mussolini's tirade against democracies and learn of yet another unwanted enemy. Allied bombing further demoralized the people. "Eighty percent of our cities were hit, and it was clear that the armed forces of the navy and air force were not as strong as the Fascist regime wanted us to believe, and so Mussolini lost all credibility in the eyes of the people," says Carla. Rome, which had not been bombed, was swollen with refugees. Shortages, confusion and fear created a rebellious mood.

The Communist Party continued to agitate. A series of labour strikes, the first since 1925 when they were outlawed, began in early 1943. Women were the first to walk off the job, and many were arrested. Marisa: "Women became very involved. There were many women in Rome who opposed the Germans en masse. There was one very famous incident when a group of women organized an attack on a bakehouse to protest the rationing of bread. Mussolini ordered police to fire into the crowds. One woman was shot and killed, and her daughter severely crippled. On her tombstone was written, "I asked only bread for my children — but I found lead."[1] Outrage led to further protest.

"Women were not even allowed to vote," says Carla. "And I think in a country where people feel oppressed, as was the case in Italy for so many years, the moment they see that there is the slightest opportunity to regain their freedom they are willing to resort to anything, even violence if necessary. We knew that demonstrations might take a violent turn, but held them anyway."

The first bombing attack on Rome in late spring 1943 further shook the people. Even those who had remained loyal to the Duce began to question the cost of this unwanted war: "Mussolini knew the mood of the people, and he didn't even go to the bombed site of San Lorenzo," recalls Carla. "He sent the Pope instead. The crowd there shouted and cried for action to stop the war."

As the "pact of steel" that formed the Axis began to crack and Italy was threatened by imminent invasion, Mussolini went to bed, where he remained for two months suffering from acute ulcers and a rare form of venereal disease. As his dream of victory and conquest in Europe turned into a nightmare, he completely lost touch with reality. The Allies landed in Corsica on 10 July 1943. Mussolini continued to rule the country by telephone, reporting to King Vittorio Emanuele II that he had an "ultimate weapon" that would ensure victory. The imagined weapon never materialized because, he said, it had to be kept top secret. The king, while maintaining the loyalty of the people and the army, had not interfered with politics since Mussolini had come to power. He had been charmed by the Duce, and believed him to be the only man capable of ending the ineffective, disorganized parliamentary system that had existed before Fascism. But now the king's confidence was severely shaken. He summoned top advisers of the government and military to report on the situation. After listening to hours of scathing criticism from respected authorities, the king finally fired the bedridden, ranting dictator on 24 July. A former assistant,

Marshal Pietro Badoglio, was appointed leader, and Mussolini was sent to an island resort where he was kept under house arrest.

Badoglio immediately began negotiations with the Allies while trying to conceal from Berlin news of Italian surrender. But on 3 September the Italian surrender was broadcast by the BBC, and German troops immediately moved to occupy Rome and two-thirds of Italy. The king and government fled the country.

In the capital, political intrigue became open aggression as those in the underground joined General Cadorna's Ariete armoured division, which was preparing to block German entry into Rome. Altogether about two thousand civilians joined the Italian division. Marisa and Carla immediately volunteered for armed action. On 8 September, only one day before the main Allied landings at Salerno in southern Italy, German troops led by Field Marshal Kesselring stormed into Rome. Near the two-thousand-year-old walls of Porto San Paolo the ragtag partisan army, hastily armed and trained by Italian soldiers, barricaded itself behind small armoured cars and prepared to stop the German advance. Marisa: "I didn't get any training. I was given a revolver and simply shown how to load and fire it. With explosives I was told to try and throw myself forward, not backwards, but nothing more specific."

The Germans approached the barricades at San Paolo in Tiger panzers. Carla: "It looked as though we were going to lose. One of the professors who had organized the action got wounded. We started to retreat, firing from behind our armoured cars. There were people everywhere, it seemed — civilians, soldiers, partisans. While retreating, one of these cars was hit by an armour-piercing shell from a German panzer, and out came a young boy, injured and bleeding. After we retreated, I took him home with me and looked after him. He was one of the first people I helped to hide and escape."

The partisan fight at San Paolo could hardly have stopped the Germans in their conquest of Rome. Declaring the capital an "open city" because of pressure from the Vatican to avoid bombing, the Germans proceeded to abuse its status by setting up their military headquarters and utilizing city streets and gardens effectively as a military depot.

The diverse resistance groups hastily began to organize united partisan activities. A former Tuscan railway worker, Ilio Barontini, who had fought in the Spanish civil war, formed and commanded a united partisan movement, which he modelled on the French maquis. Within the organization he created a small, tightly knit group of fighters, mostly young intellectuals, he called Gappisti (after their acronym, GAP — Groups of Partisan Action), who were directed to disrupt German occupation forces and leadership. Both Marisa and Carla became Gappisti. Marisa: "I asked to be armed. I wanted to participate. I thought that armed resistance was the best and simplest solution."

Carla says, "GAP was the armed section of the whole conspiracy. There were other sections — those who took care of disbanded soldiers and helped Allied prisoners to hide and escape, a political section involved in keeping the various groups co-ordinated, and then the armed section. We divided the city into eight zones in which Gappisti operated.

"Our first problems were inexperienced saboteurs and a shortage of materials. The first steel incendiary bombs were smuggled out of the Romano Gas Company by some of the employees and were sealed with plaster, which is not very strong, and so the explosion was not very powerful. Our sabotage expert was arrested immediately, and so it became necessary for us to get involved in the preparation of explosives. We got hold of a nuclear physicist, a university professor, and his wife and used the cellar of a building located near the Colosseum. It was there that I had my 'course' on arms. Naturally it was impossible to get any military training whatsoever in the city."

Marisa: "I was told that some were wick-fused explosives, and I only had thirty seconds to get away, and that others were acid bombs, and they were to be turned over. But on the whole we got no training of any sort. We were given bombs and told where to place them in different locations, and that was it.

"Girls were particularly useful in the armed resistance," she adds. "I'm talking about urban war, in a city, so it was easier for a girl to go unnoticed. If a girl was caught with arms, that would have been extremely dangerous — it would mean her death. But some specific actions were successful precisely because girls were under less suspicion. You have to keep in mind that whereas in today's terrorism the figure of a woman shooting is very common, forty years ago it wasn't regarded as possible or probable that a woman could shoot. One didn't perceive the possibility and this became a great advantage for the partisans."

"That's true," says Carla. "Naturally the Germans didn't think that a woman could have carried a bomb, so this became the women's task. Women always carried the bombs until the moment of action. And some, like me, set them off as well. But in many instances women were not given arms because men believed that they were more emotional and less capable of making decisions. Personally, I have always believed that anybody can do it but that each individual has to know his or her limits, and the extent that he or she can count on having the moral strength — to ask oneself what would happen if one were arrested and tortured and if one would be able to resist without betraying one's comrades. I was convinced that I did possess this inner strength." But Carla admits that despite her personal convictions, in the beginning she had to fight to be treated as an equal: "When I first joined I had a gun of my father's from the 1914-18 war. Another partisan, Rosario Bentivegna, and some other men took it away from me. However, I was riding on the bus one day and happened to be standing close to a German soldier and I stole his gun, a Beretta sub-machine-gun. So that is how I regained a gun. Eventually the men accepted us, but it was difficult in the beginning. Many women were directed to work in the underground as couriers and so on. Others were trained in medicine because we were prepared for the city to rise up, which never happened. Men couldn't tolerate the idea that an action might not be successful because a woman had been involved. So there were some problems. Women had to assert themselves. I was finally allowed to participate in difficult actions because of my strong will. The same happened with several other women."

Marisa had the additional disadvantage of being so young.

But her shy, sweet looks were deceiving. She convinced her commander to let her participate. "Since then I have asked myself if the men treated us as equals. At the time I never questioned it. I think they did, but I don't remember clearly. During that period we were a small group of intellectuals, and it seemed to us that we were all equal. One thing is for sure — Italian men know how far they can push us before we put an end to it!" she laughs. "Women are probably less violent than men, but maybe because of that, when we are convinced what we are doing is right, we act with great conviction and determination."

After German take-over and reinstatement of Mussolini as a puppet leader, Italian Fascists sporting their black shirts and daggers again were seen in public. Rome was swollen beyond capacity. There were fifteen hundred German soldiers in the centre of the city at any one time, unknown numbers of refugees, and cellars full of escaped Allied prisoners, Italian soldiers and Jews trying to make their way from the Vatican neutral zone through the underground to safety. The Gappisti had to plan each action carefully in the overcrowded city. "We had to study each detail to guarantee maximum safety to women and children and old people," says Carla. "And we had to be careful not to be recognized, which was difficult since Rome was our home, where we had gone to school, so many people knew us. We ran the constant risk of being recognized. It was often difficult to find shelter — you literally stumbled on Germans and Fascists every few yards." After the Gappisti assassinated two Black Shirts, "they began wearing trench coats over their uniforms to avoid being subject to our attacks."

Marisa and Carla participated in one of their first armed actions together on 18 December 1943. Carla: "There were four of us, two men, Marisa and myself. We were planning to attack the Fascist commander in Rome, Pizzirani, who lived on Via Keren, a street that crosses Via Nomentana. We studied the details very carefully. We had identified his house and everything.

"My task was to block the car, and the others were to shoot at Pizzirani and his secretary and the driver. I had not been allowed to carry a gun, because they thought I was too exposed as the one stopping the car and I would be identified. Also, they thought that being the one to stand on the road with the car coming straight at me, I would panic and shoot.''

When Pizzirani arrived in his car, Carla rushed out to stop it. Marisa and the others ran out and began to shoot, but all three guns jammed. By the time they had taken out the damaged bullets and replaced them, Pizzirani had escaped unharmed. Carla: ''Pizzirani got out right next to me. If I had had a gun, I could have killed him. Unfortunately, only the secretary was severely wounded, and our mission had failed.''

Carla's first assignment to assassinate a German officer came after several other Gappisti had failed. She planned to wait for him outside the Hotel Excelsior, where the Germans had set up headquarters. The hotel entrance was blocked by barbed wire and a road check. ''He carried important documents regarding the defence of Rome, and it was imperative I get hold of his briefcase,'' she explains. ''With three others covering me, I waited until he came out of the hotel. I hesitated till he had passed the police, and when there was no one around, for the first time I fired against a man.

''It was a traumatic experience. I almost wanted to call him, to make him turn around . . . but I knew he was armed. It seemed impossible that with my peaceful disposition, against any form of violence, I should hold the gun, point it at him and shoot him in the back. I took his briefcase. I was in shock. I had the briefcase under my arm and I began running down the street with the gun still in my hand, without thinking of putting it away. It was raining and tears were streaming down my face. He didn't die right away, but lay there moaning. His moaning was torture to me, all through my flight down Piazza Barberini.

''Rosario Bentivegna then joined me and tried to calm me down. He told me that the first time he had to shoot a man he felt the same way, and that I should not feel ashamed to feel a sense of shock that results from that first contact with violence and brutality.

''After getting over the initial shock, especially since many of our comrades were being arrested and tortured, all our scruples were replaced by sheer determination to fight for our cause. So in a way this sense of guilt, or any feeling that our personalities were being brutalized, disappeared. The fight was becoming more intense, very fierce, and daily we were learning to hate the Germans more as one by one our friends were killed."

Marisa: "They say that women have babies, and so don't kill. I was very young and very determined. At that time it was clear that each Nazi I killed, each bomb I helped to explode, shortened the length of the war and saved the lives of all women and children. The only feeling I had at the moment was worrying about being caught. There wasn't anything else going on in my mind. I never asked myself if the soldier or SS man I killed had a wife or children. I never thought about it. In those situations you just don't think about anything but surviving and covering for your friends to help them survive.

"My parents helped me a great deal," she continues. "They gave me strength. They knew about what I was doing — not specifics, which would have been unwise to tell them, but they knew I was engaged in armed combat."

On 18 January 1944 Carla and Bentivegna participated in an attack on Rome's main prison, Regina Coeli. Several other attempts had been made, but all had failed. Carla carried the bomb to the top of an elevated embankment immediately above the entrance to the jail. The Gappisti timed it so that an army truck full of German guards would be outside the entrance and would be killed by the bomb blast.

"I passed the bomb to Mario Fiorentini, who lit the fuse and threw it. He fled on a bicycle. I was armed with a revolver to prevent him from getting stopped during his flight. The Germans immediately climbed up the embankment and started shooting, but fortunately he escaped unharmed. He got to the other side of the Tiber River and then disappeared in the little lanes of the area. I had to flee to avoid being shot. Naturally the Germans didn't shoot at me. They didn't think a woman would have been involved."

Carla and Bentivegna often operated together in the area of the Centro Storico, where German military supplies were kept. On one occasion they participated in a bombing attack on an opera house in which fifteen hundred German soldiers were being entertained. Sixteen were shot as they rushed out of the building. Carla was identified on 7 March after she single-handedly blew up a German fuel truck at a depot in Via Claudio.

"I went there at eight o'clock in the morning," she says. "There were children going to school so I had to wait before placing the bomb in a large truck that was parked in front of the depot. I stood around for about half an hour. A couple of German soldiers noticed me and they began flirting, offering me a cigarette. After a while they resumed their work, loading their truck, and paid no more attention. So I lit the fuse, placed the bomb in the truck and began walking away. There was a huge explosion. First the drums on the truck exploded, and then the depot. The fires lasted the whole day, and my friends told me later they heard the roar of the explosion half-way across Rome." Twenty-five hundred gallons of gasoline had exploded. The walls are still blackened today.

"From then on I was on the wanted list. There was a heavy price on my head. I carried no identification papers with me, so I certainly was a good candidate for getting shot or at least tortured. But I was lucky. One time it was Marisa Musu who saved my life."

Both women described the incident, but Marisa was typically reticent about her own role in it. She said, "The Germans had rounded up seven thousand men to send to Germany for forced labour. All their relatives marched to the prison barracks where they were held and demanded their freedom. Literally thousands of women marched there. One woman, Teresa Gullace, had brought her husband a piece of bread. When she tried to hand it to him she was shot and killed by a German soldier."

"I saw it happen," says Carla. "I was about to draw my gun, to kill the soldier who shot her, when a Fascist policeman grabbed me and pulled me into the police headquarters. Luckily, the other women tried to prevent them from taking me and were pulling and tugging, and in the general chaos Marisa was able to get close and take my gun from me. When I got inside I was searched. Miraculously, they pulled a Fascist ID card from my pocket. It had her name, and no photograph. It saved my life. At that time she still wasn't wanted by the police, but if they had found out my true identity, they would have killed me.

Courtesy Diana Barnato Walker

Diana Barnato Walker

Courtesy Joan Cowey

Joan Cowey

Courtesy Diana Barnato Walker

Diana Barnato Walker

Imperial War Museum, London

ATS women on AA gun site

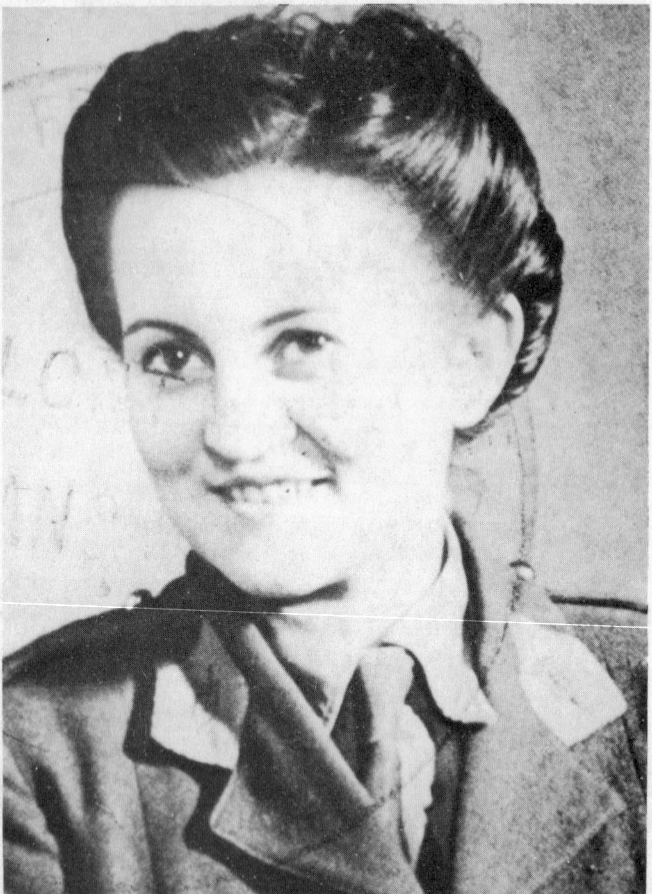

Courtesy Jeanne Bohec

Jeanne Bohec

Miller Services, Toronto

Inmates of Bergen-Belsen

Miller Services, Toronto

A European partisan, her name and nationality unknown, captured and killed by the Nazis

Courtesy Irena Komorowski

Irena (''Black Barbara'') Komorowski and fellow resistance fighters, Warsaw uprising

Courtesy Ida Kasprzak

Ida Kasprzak

Wojskowy Historical Institute

Home Army blockade, Warsaw uprising

Wojskowy Historical Institute

Soldiers of the Emilia Plater Battalion

Courtesy Sophia Kuntsevich

Sophia Kuntsevich

Novosti

Katyusha Mikhaylova dressing the wound of a soldier

I began fabricating a story, saying I was a Fascist who had gone there to deter the women from protesting and that, by mistake, I had been taken for a subversive. So they let me go."

Musu, still unidentified, continued living at home most of the time, but Capponi left her mother's apartment and began living underground. She often spent her days carrying out GAP actions in Rome and then travelling to Centocelle, a suburb where the Germans had a number of barracks, to operate at night. "During the day I was a Gappista, and at night a member of another GAP organization in Centocelle where we attacked the columns of German soldiers. This was the most difficult period of time for me. It was very stressful because of the constant travelling in the incredible cold, inadequately dressed, inadequately fed. I slept very little and was constantly hunted. In Rome I slept in a cellar, and at Centocelle I slept on a floor without a mattress. At dawn, right after the curfew, I had to go back to Rome on foot." Bicycles had been outlawed, and the Gappisti seldom had money for trams.

"Often I slept on a floor in someone's cellar among prisoners who were trying to escape — Englishmen, Canadians, Americans, Poles. The people were wonderful. There was a tacit conspiracy against the Germans. In fact Roman women are credited with helping 640,000 foreign POWs escape to safety.

"But still, this kind of life left a scar. It was very, very hard, sleeping on cellar floors with fifteen to sixteen men, dirty, cold, infested with lice. I suffered from scabies and other diseases that eventually resulted in the loss of one of my lungs. But it didn't diminish my morale; I think it developed character."

Rosario Bentivegna, who later wrote about those days in his book *Achtung Banditen!*, recounted an incident in which Carla was followed towards Centocelle by an Italian fireman. She was angry, thinking he wanted to talk or flirt, and walked quickly to discourage him. Then they arrived at a road block and were stopped. Carla was carrying explosives in her bag. The fireman whispered something to the German police, and they let her go without checking her. She waited to thank the fireman. He told her that he knew of her and her activities and had wanted to show his appreciation by helping her. Carla thanked the man. "I should be the one to thank you and the people like you who are fighting to free us from these assassins," he said. Then he added, smiling, "Do you know what they call you here? They call you the little English girl, but you are more Roman than I am!"[2]

Bentivegna wrote, "Carla's presence among the partisans was very much felt and it could not have been otherwise. She covered herself in heavy wool clothes, knee-high wool socks and a pair of military boots given to her by Enzo Russo (another partisan), and soon in Centocelle was born the legend of this young blonde woman who went out at night to shoot Germans."[3]

Hopes for a quick Allied victory in Italy were dying. On 16 February 1944 Allied forces landed at Anzio, but German resistance was fierce and fighting continued for three days. Morale was poor in the capital, where constant bombing disrupted public transport, water works and electricity. There were severe food shortages and little heat. The partisans were weakening. They could afford no more than one uncooked meal a day. The British SOE and American OSS secret services sent fifty agents into Italy but were hesitant about giving much assistance to the largely leftist partisans. Churchill was adamant that no partisan uprising take place in the capital. He wanted no repeat of the partisan civil war that was raging in Yugoslavia, in which partisan armies on the right and the left spent as much time fighting each other as they did the Germans. Further, London and Washington did not want the Italians to exchange Fascism for Communist rule after the Germans were defeated.

Terrorist attacks by the partisans demoralized German troops and infuriated German command. When partisans were caught their treatment was barbaric, and fear of capture had become a constant, immovable force hanging over all their lives. Fear had become as painful as hunger and fatigue — perhaps more. Carla: "If caught, we knew we would be exposed to great suffering. For example, two women partisans were thrown to the police dogs in the headquarters of the Fascist militia and torn to pieces as a last torture for refusing to talk. A horrible death! The best-known example was Irma Bandiera of Bologna. She was a courier between the city and the mountain, and she knew all the names of the men in charge of the partisan groups and all the locations from which they operated. She was tortured repeatedly for days and days. They pulled her fingernails out, disfigured her face and even tore off a breast. And then, when she died, they threw her body in a rubbish dump and prevented her mother from retrieving it for burial for three days.'' The news of friends' torture and deaths hung over them. There was sadness, horror and foreboding.

The German High Command in Rome discussed other means of contending with the partisans. They decided to bring in a specially trained unit to conduct search-and-destroy missions in and just outside Rome. The unit was the 11th Company of the 3rd Bozen Battalion, which consisted of 156 men. They arrived in Rome and began training each day in the city. GAP observed the unit and noted the time each day when the column marched through the centre of the city as a show of force.

GAP commanders decided on an action that would demonstrate to the Germans that they were not intimidated and that they hoped would raise morale among the population. They planned to attack the unit as it crossed the centre of the city one afternoon.

''This episode would be central to the resistance movement. It would be our statement and we knew it had to succeed,'' says Marisa.

''They had arrested seventy percent of the partisans in Rome, and forty percent had been executed,'' says Carla. ''We all knew we were identified. I was nervous and upset during the planning for Via Rasella.''

The original plan was to attack the soldiers as they marched through the Piazza di Spagna (Spanish Steps), the beautiful square beneath the cathedral where, in better times, Blake and Shelley had lodged and composed poetry. GAP changed its plan after deciding that it was too risky for civilians and chose instead to ambush the column as it passed down a narrow cobble-stoned street that ended at a tunnel, named Via Rasella. The attack was set for 23 March 1944.

''Someone thought it would be a good idea to place the explosives in a rubbish cart, the kind used by the Sanitation Department,'' remembers Carla. ''Then it was a matter of finding a garbage collector sympathetic to us. We found a candidate, but I felt that he wasn't right. He showed signs of weakness, and I always trusted my instincts in these things. We didn't use him, and later he betrayed us to the SS.

"I suggested Bentivegna as the garbage man. I worked with him often and he appeared to be determined and courageous.

"It was a complex plan. We used two GAP sections. Bentivegna was assigned the task of placing the explosives and I had to act as his cover. I was supposed to stay in front of a nearby building until I received a signal. This would come once the others heard the Germans enter Piazza di Spagna. I would then proceed toward my position on the corner of the street, go up to Bentivegna and give him a signal. He was to light the fuse."

"We waited at another spot to throw three mortar bombs as well and cover for the others," says Marisa.

"But there was a delay," Carla recalls. "The Germans who passed by every day at the exact same time were over one hour late. We almost decided to give up. Bentivegna, dressed as a street cleaner, had pushed the rubbish cart with explosives a long way, and not without great danger of being stopped by other garbage men or the police. I had already been stopped twice by inquisitive policemen. They might have wondered why I carried a man's raincoat on a perfectly cloudless day. It was becoming very suspenseful. At a certain point I couldn't see Bentivegna because of the marching soldiers, and at that point I walked toward our contact point and gave him the signal. I handed him the trench coat for his get-away.

"He lit the fuse, and we began to run towards Via Nazionale. The other partisans threw mortar bombs."

Thirty-three men were blown to pieces and another seventy were wounded. They lay amid pools of blood and chunks of concrete, shattered glass and pieces of bodies. The whole area of the city was in chaos. Within minutes truckloads of Germans arrived and began shooting in the streets. Rumours spread that hundreds of people were being shot. SS men began pounding down doors, going from house to house and dragging the inhabitants into the streets. They lined up all the men they found in the area, and with hands forced above their heads the innocent civilians were taken away.[4]

"That same night," says Marisa, "the Germans took 330 prisoners from jail — many were the people they had just pulled out of their homes — and they took them to the Ardeatine caves."

Carla and Bentivegna rushed to her mother's apartment, where Bentivegna fainted. Her mother revived him with some medicine before the two made their way to a hiding place.

On orders direct from the Fuehrer himself, German commanders in Rome ordered the death of ten Romans for every German killed at Via Rasella. The prisoners were rounded up from jails throughout the city. Some were partisans; most were not. With hands tied behind their backs they were herded into trucks. Any who did not suspect their fate immediately knew it when the trucks screeched into the grounds surrounding the Ardeatine catacombs. German soldiers, numb with alcohol, had lit torches inside the caves. Then, five at a time, the prisoners were led deep inside, forced to kneel against the walls, and were shot in the back of the neck. Night became dawn as the massacre continued, the living forced to climb onto the pile of dead. The Germans sealed up the caves when the night's work was done, and set off explosives.

By mistake, 335 prisoners were massacred — 5 men too many. The German commander, a man who hated inaccuracy, was furious.

The following morning, 25 March, a 117-word communiqué written by the German High Command was posted throughout Rome. It announced that in reprisal for the murder of German soldiers, 330 Romans, ten for every German killed, had been shot. In what has become the most controversial and haunting episode in the history of the resistance in Italy, Roman citizens, in sorrow and in rage, began to take sides. The Pope himself, whose own war record remains controversial,[5] condemned the partisans: "Thirty-two [sic] victims on the one hand and on the other three hundred and twenty persons [sic] sacrificed for the guilty parties who escaped arrest." He then called on the people of Rome to stop the violent actions against the Germans.

Other factions of the resistance were divided about the Via Rasella attack. An American OSS agent in Rome, Peter Tompkins, wrote, "Now there was no telling what the Germans' reaction would be: certainly it boded no good for the underground in the city."[6]

When Carla learned of the reprisal that morning, she felt "anguish. There was and still is so much bitterness and hatred generated by the Ardeatine massacres in reprisal. Twenty-two of those killed there were partisans who had been imprisoned at Regina Coeli. They were our friends. People blamed us; they wondered why we didn't turn ourselves in to prevent the massacre. But nobody ever proposed to us the possibility of preventing a massacre by giving ourselves up. As a matter of fact, the Germans proceeded very secretly for fear that the partisans would learn of their plans and attack the trucks which took those 335 people to their deaths.

"We were even criticized by some anti-Fascist groups," she continues. "The partisan group had two exponents: those who believed in organizing a plan of action but who preferred to wait for the ideal opportunity, and those who believed in immediate action. We were convinced that it was necessary to make life difficult for the enemy and in this we had the full support of the Allies in our various attacks and assassination attempts . . . very often paying dearly for it with our own lives."

"The reprisal made me realize even more strongly that the Germans had to be stopped," says Marisa. "I never considered leaving GAP."

Of the forty-six operations carried out by GAP during the ten-month German occupation of Rome, only three occurred after Via Rasella. Historians agree that the attack did not achieve its aim of keeping up morale or the momentum of the resistance movement. The civilians, whomever they blamed, were terrified and sickened after the massacre.

One of the last GAP actions in Rome was an aborted attempt to assassinate Benito Mussolini's son, Vittorio, who had dabbled in politics. Marisa Musu and two male Gappisti planned to sneak into his house at night.

"We planned the mission very carefully," says Marisa. "But when we arrived there, policemen were waiting and opened fire. We were caught.

"What had happened was that there had been a theft in the house across from Mussolini's the night before. The police were actually lying in wait for the thieves to return. They caught us instead. We were the first in our unit to fall into the hands of the police. Luckily they believed that they had the thieves. Under martial law, anyone found carrying arms was to be shot on the spot, but we were put in jail as robbers. In the midst of all this there was a traitor who learned we had been arrested. He had been involved with GAP and knew all about our activities. He went to my parents' house and told them that if they didn't pay him a sum of money, he would report me as a partisan. My father, a very peaceful man, took out a knife from his desk and said to the informer, 'Get out of here or I'll kill you.' He didn't think, 'I'll save my daughter by paying him anything.' I have always been very proud of him for this. So anyway, the informer disclosed our names and told the police we had been participants in the ambush on Via Rasella.

"Our GAP commander decided to do everything possible to help us escape from jail. The group devised a well-detailed plan according to which I was to pretend to be suffering from an appendix attack. The chief of the jail was a supporter of the resistance and knew of this plan. I was to go by ambulance to a certain hospital where there would be several doctor-partisans who were supposed to pretend to operate on me. Then, during the night, they would help me escape. The whole situation became entangled because I made a mistake on the date that this was supposed to take place. I got the message in a thermos. The coffee had leaked through and the date wasn't legible. So I started the escape one day too soon, and of course the doctors who were aware of the situation were not there. Since I was only a prisoner they said that they had no room, and I was sent to another hospital where no one knew anything about the plan. Unfortunately, because they had orders to operate immediately, I was taken directly to surgery.

"At this point I decided to risk it. I was very scared and alarmed. I thought, 'They are going to operate on me, and then tomorrow the Germans will take me, and finding me in this weakened condition, it will be easier for them to torture me.' So I decided to chance it. I told my story to the head surgeon, counting on some sympathy towards the partisans. He told me not to worry. He said he would take care of it. I was sent back to the ward 'for examinations.' The whole thing turned out to be quite funny. I stayed there two days, and when the surgeon came on his rounds with the interns he asked them to examine me. Naturally I gave signs of discomfort at inappropriate times. The surgeon asked the young doctors what they thought the diagnosis was, and they would come up with the most absurd possibilities.'' She laughs. ''I wonder how many medical careers were ruined?

''In the meantime the members of the resistance intervened. Since I was in a room guarded by a policeman, the partisans talked to him and convinced him to flee with us because the Americans were nearby now, and it would be advantageous for him to do so. That night, the policeman and I escaped. I was taken to a convent, where I was when the Americans entered Rome.''

Carla and Bentivegna had narrowly escaped capture. They had been offered the use of a house in Rome after Via Rasella, but they were betrayed by its owners. ''Not long after we had taken shelter there, one night after curfew — there were four of us — the Germans came and tried to break down the door. I kept them from breaking in, telling them I was alone and getting dressed. Meanwhile the other partisans jumped out the window. Bentivegna didn't jump because he wanted to wait for me. I took the safety pin out of a time bomb, and got ready to jump as well. I looked down and saw that the others had twisted their ankles or hurt themselves in the fall. So, being a woman — women are more reflective — I saw there was a drainage pipe on the side of the house. I threw the bomb, and slid down it. So there I was calmly getting down when I realized I had no shoes on! We escaped and Bentivegna took us to a friend's home. She let us in. I had no shoes and was wearing a beige sweater that had got very dirty from sliding down in the grass — you could see that I was on the run. Anyway, she was very kind as well as discreet. She offered me a glass of milk, an exceptional thing for me. I hadn't had a glass of milk for months. And she gave me a pair of slippers.

''The next morning around five o'clock I went out in the slippers. Life could be comic as well as tragic! I had to warn the others about the betrayal. On the tram a man and a woman approached me and whispered, 'Do you need any help?' They had seen by my clothes, and the fact that I had no coat in this terribly cold weather, that I was on the run. They had also seen that I had no money for the fare and the driver had let me slip by without paying because there was a tacit conspiracy against the Germans.

"After this escape from the house, GAP decided to send us to another area outside Rome. They sent us all away, because we were identified, and got a whole new group of people."

Carla was sent to the tiny village of Palestrina, about one hour by car from Rome, between Rome and the Allied-German front. It was a simple, poor agricultural village, but a famous architectural site. A Renaissance palace, Castel San Pietro, had been built on the site of an ancient Roman temple, which was terraced down a hill. The Germans had set up headquarters in the palace, and there were German regiments stationed nearby. Most of the young men of the village had fled to the mountains to avoid German labour drafts, and a group of about 150 partisans was active there. Most of the peasants of the area either helped or joined the partisans.

They ambushed and attacked German units each night. Eventually they overran an army kitchen and infirmary, capturing forty-seven German prisoners whom they kept in a cave in their mountain hide-out. Despite severe reprisals the peasants remained loyal to the partisans.

Carla: "One of our first actions was to free some Soviet prisoners who were held in a nearby jail. They had been recruited from the concentration camps in Germany by the Germans and had agreed to work for them to escape the gas chambers. They were used to clear mines at the German front. Poor people. . . . They were used to clear the mines by walking in a line with linked arms, and at night they were intoxicated with liquor provided by the Germans so as to forget what they were made to do. We helped them escape from the jail where they were kept at night. A dozen of them managed to escape but two or three were killed. They were very brave even though they tricked us from time to time. For example, we would give them the signal and wait for them to do their part, but they would have disappeared somewhere."

Palestrina was very near the Allied-German front, and Carla says it was bombed from both sides. But the experience of living among the peasants was memorable. "It was here that I came across peasant women who behaved with extraordinary heroism. Entire families got involved in partisan actions, and the women would feed us even when they were practically starving themselves. One family, the Pincis, were captured and killed by the Germans. We later found them raped and tortured beyond recognition. Their mother had been tied kneeling down to witness their death. She later went mad with grief.

"In one instance some peasants called us for help because the Germans were ransacking their farmhouse. So seven of us went down from the mountain and found the Germans just finished eating and drinking, lying down in the wheat fields. We approached them silently, but startled them, and one by one they jumped up. We pointed our guns at them and led them along in front of us as prisoners. I had two soldiers whom I had disarmed in front of me. They were petrified. At a certain point, though, I noticed that their expression had changed — now they seemed happy. They had noticed another German soldier had got up behind me and was about to kill me with his machine-gun. But one of the partisans saved my life. He fired against the soldier. This started a series of shootings and for about half an hour we fought non-stop. We must have killed about ten of them. We took all their arms but soon had to retreat as they began to get reinforcements.

"We had captured many prisoners, and I found myself forced to defend them against the local partisans who wanted to kill them. 'Why should we have to feed assassins?' they would demand. 'We can't keep them.' So I tried to convince them with a political argument. I really detest violence and abusive behaviour, so I pointed out that these prisoners would come in handy at a time of negotiations. I also pointed out to the partisans that we had not only captured them but at the same time we had got hold of an army kitchen near Palestrina which provided food for the entire population and an infirmary with a doctor and nurse. So we had officers who would be precious candidates for exchanges with the enemy, and I was adamant they not be killed. I got my way. We kept them in a cave that had been used by shepherds as a fold for the sheep — not the most sanitary conditions. The local peasants voluntarily brought us food — lambs that had been hastily boiled in water with all the internal organs, without salt. . . . I must say that I ate very little. And it was hard work just keeping prisoners alive. They had to be fed and taken out of the cave periodically, and this was hard work with forty-seven men. In a way the partisans who wanted them dead were right, because it was even dangerous. If any of them escaped it would have caused a lot of trouble. As the number grew it got even more dangerous to keep them."

Carla and Bentivegna's feeling for each other had grown stronger since the Via Rasella incident, particularly after the condemnation to which they had been subjected. "It was only natural, then, that being together evoked feelings of mutual respect and also, at times, of love. We worked together very closely. The partisan commander always put us together without thinking of the possible sentimental repercussions. The mere fact that we were living through such dramatic and dangerous moments together made us more protective and sympathetic toward one another. On several occasions I had saved Bentivegna's life.

"I do think those dangerous moments we shared brought us closer together. To tell the truth, in the beginning I found him quite disagreeable because he always wanted to do everything himself. He wouldn't even let me have a gun. But as time went on . . .

"We really discovered a deeper feeling than just friendship or affection when we first left Rome on foot to operate with partisans in the countryside. During this journey we sometimes laughed, sometimes . . . felt freer than we had in a long time. A disbanded soldier had joined us. We were walking along the railroad that was no longer in use. At a certain point, almost from nowhere, American troops started to machine-gun us, possibly thinking we were the enemy. We threw ourselves down to the ground, knapsacks, bikes and all. I had fallen on my knapsack with the explosives in it. The soldier, realizing that the railroad was not a safe place, ran away. We were machine-gunned again and a tree fell on us. So there we were, under the greenery of the tree, and I think it was the first time that I kissed Bentivegna."

By the first day of June 1944 the Germans were retreating in areas around Rome. As the American troops approached the city the partisans called upon the people for a general uprising, something that had occurred in other major cities such as Naples. (The Romans would disappoint them.) Bentivegna and Carla were ordered back to Rome to help in any uprising that might occur.

"We were told through a radio-transmitter to go back to Rome to prepare for the upcoming insurrection. There was a cannonade and the Allied shells prevented us from moving at all because of the constant falling of bombs, shrapnel and so on. It was hell. I was hit by one of the splinters in the hand and in the back, but they were only superficial wounds. More than anything else I had many splinters in my face, around my lips. It was during this barrage bombing that we received the order to return to Rome. We decided to go there on foot. We had a couple of combat encounters with Germans along the way.

"When we got to the outskirts of Rome we were given knapsacks with explosives. One of the partisans who was with us had been severely wounded in his leg, so he couldn't ride a bicycle. I had been injured in my knee but it was less serious so I took the bike and together with Bentivegna went towards Tivoli. But when we got to Ponte Marmo, on the road that leads to Tivoli, we came across Germans with tanks who took away our bikes. Luckily they didn't look into our knapsacks or we would have been shot on the spot. Then we walked through the fields, and there we met the Americans who told us not to go back to Rome because the insurrection would not take place."

On 4 June 1944 the Americans entered Rome. Marisa, still hidden in a convent, returned home to her parents. Carla and Bentivegna returned to Rome and continued their work with GAP. "We resumed our normal active lives. We occupied the newspaper offices and printed our first newspaper after the liberation with big banners — Long live the Americans! Long live the British! Long live the Canadians! Long live the Allies! But it was more than a year later that the last Germans capitulated in Italy. They continued to hold on in areas . . . and we continued the fight. By the end of the war, exhausted and ill, I weighed only 39 kg [eighty-six pounds]. . . .

"The fact that many of us ended up marrying other partisans — Marisa Musu, for example married her GAP commander — has always been a favourite teasing subject, especially since all our friends and acquaintances were partisans. I didn't have any Fascist friends! And I was substantially changed, so much more mature, which shooting a gun makes you. I never danced, because while I was young there was a war on, and even after the liberation of Rome it continued. I matured prematurely, and I became more responsible. I married Bentivegna in September. He was about to be parachuted into Yugoslavia to fight with the partisans there. I married him because he said to me, 'Perhaps I will die. . . . Maybe I won't come back from this second partisan activity. I don't know when it will end. I want to have a child.' The whole thing seemed so romantic. I told him, 'We don't have to get married to have a child,' but no, he was concerned about the Italian laws of the time. He couldn't give his last name to the child unless we were married."

On 20 September at seven o'clock in the evening Carla and Bentivegna were married in St Mary's Church in Campitelli. Almost all their GAP friends were there, except those who were now fighting in northern Italy. One of Carla's friends lent her a pair of shoes for the wedding. Bentivegna described the ceremony: "It was raining hard and there was no light. Suddenly the light came on for a few seconds. We thought it was a good omen but later, during the night, there were only a few candles lighting the hall. Caronia [a friend] and his wife wanted to celebrate in some way. They took us in their car to their house where they had prepared a real dinner and we also had a bottle of sparkling wine."[7] The next morning at seven o'clock Bentivegna left by truck with the Allies for Bari, from where he was parachuted into Yugoslavia.

Marisa remembers the readjustment to normal life in Rome: "Those first months were difficult. Getting back to normal wasn't easy. First of all, I think, it is very hard to live without a gun once you have had one. A gun gives one a great sense of security. So the temptation to continue to carry one lasted many months. The most difficult thing was to go on with life as normal."

Field Marshal Kesselring later estimated that the Italian partisans had been responsible for the deaths of 13,000 German soldiers, and at least that many again were wounded. Of the several hundred thousand partisans who fought throughout Italy, it has been estimated that 35,000 were women, 4,653 of whom were arrested, tortured and beaten to death, 623 killed or wounded in action, 2,750 deported and 15 awarded the Gold Medal for Military Valour. Most of the women who were honoured with the Gold Medal received posthumous citations.[8] Marisa Musu received the Silver Medal for Military Valour. Carla Capponi was promoted to vice-commander by GAP and received the Gold Medal before a full military review in 1949. Her husband, now a physician, held their small daughter Elena and watched her receive the tribute. Her citation read, "With gun in hand, first among the first, she distinguished herself by her spirit of sacrifice towards her comrades in the face of danger."

But the four years since the war ended had not been easy for any of the former Gappisti. In 1948 a small group of relatives of the Ardeatine victims filed a legal suit against some of the partisans of Rome. Carla and Bentivegna, along with five others, were named as defendants.

Despite the Roman court's decision that Via Rasella was a legitimate act of war in which none of the partisans was answerable for the following reprisal, public criticism was never fully silenced. Most of the Gappisti were honoured and had gained prestigious positions in post-war Italy, but still there were those who held them accountable. Post-war politics in the country continued to be divisive and heated, and Carla's election to Parliament in 1953 as a deputy of the Communist Party kept her in the public eye. She continued to receive obscene phone calls and remained the target of some abuse. She says, "It was personally very upsetting. We were all attacked, but myself in particular. We were even threatened by the Fascists." In 1954 the group of relatives took their case to the Roman Court of Appeals, and again it was dismissed, the judgment citing the action "as an act of war, sanctioned by the State. . . . There can be no guilt of any kind."[9]

Carla and Rosario Bentivegna were divorced a few years later because "our personalities were so different," but have remained close friends.

"You asked me if I would do it again," she smiles. "Yes, certainly. I don't know if, at my age, I could do the same things I did then, but I would certainly try."

Marisa undertook a career in journalism, covering wars in Mozambique and Vietnam. She says, "Today's pacifist ideas make me reflect more, but if I had to, I would kill again every German I killed. I still believe it was the right thing to do. My opinion has not changed."

UPRISING
Poland, 1939-45

I was taking a message to headquarters. I had the papers hidden in my sleeve, and I walked down the street pretending to be an elegant young lady out for a stroll. Then I passed by some SS men. I thought, "You bloody bastards. You think you are so strong and I am so weak, but my work will eventually defeat you." That was my satisfaction.

Black Barbara,
Courier, Home Army, Warsaw

I think it is about time that people took notice. It should be said, once and for all, that women fought too. Men believe if you don't shoot or carry a gun, you are not a soldier. But we did the most tedious and most dangerous jobs. Women were injured and killed in action. Men just don't want to admit that we fought too.

Ida Dobrzanska Kasprzak,
Officer, Home Army, Warsaw

During the war in Poland hundreds of thousands of women joined or aided the resistance.

Many of the Jewish fighters in the Warsaw Ghetto Uprising in 1943 were women. Jewish partisans, hiding and operating in the forests also appear to have had a large proportion of women in their ranks.

There were, in addition, a large percentage of women among the Communist partisans. Many went to the Soviet Union, where they received training and joined Polish regiments formed there, in much the same capacities as the Russian women.[1] In one all-female regiment, three hundred young women operated in the forests of eastern Poland, fighting entirely without men. They named their regiment after a famous Polish heroine of the eighteenth century, Emilia Platter.

In England Poles who had escaped joined General Wladyslaw Sikorski and formed a Polish army in support of the Allies. A women's auxiliary was formed, which would number 6,700 women. It functioned, like the Free French, in co-operation with the British Auxiliary Territorial Services.

Finally, forty thousand women were sworn-in members of the Home Army in Poland, led by former army officers who recognized the leadership of Sikorski in England. Thirty-five thousand were given military training in the Home Army's forest camps and functioned in all capacities, including sabotage, weapons building, message carrying, liaison, weapons smuggling, assassination and armed fighting. The leader of the Home Army in Warsaw during the city's uprising in the late summer of 1944, General Bor-Komorowski, would estimate that of his forty thousand soldiers in Warsaw, one-seventh were women.

In addition, unknown numbers of women risked their lives to help the resisters, to offer them food, hide them in their cellars and keep the silence. Girl Guides, perhaps the most exceptional example of all, scouted escape and supply routes through the city sewers, smuggled weapons, participated in assassination squads and joined armed combat in the uprising.

It is interesting to note that despite the overwhelming participation by women, and their military training, in the final analysis most men did not approve of women's bearing arms, and because there was always a shortage of guns women were the last to receive them. Furthermore, despite the equally if not more dangerous tasks women were assigned, many Polish male veterans today do not acknowledge that women were soldiers, precisely because they did not always bear arms.

Two of Warsaw's most dedicated Home Army "soldiers" were Irena Kwiatkowska Komorowski ("Black Barbara") and Ida Dobrzanska Kasprzak. They did not meet each other until they were interned in Bergen-Belsen concentration camp in Germany, in the last months of the war, but they remain friends today in Toronto, where they both live.

Ida Kasprzak is a stately woman of generous proportions, attractive face and brisk, warm manner. It is difficult to imagine the young women of twenty-one who weighed ninety-six pounds at the end of the war. She claims she has been a leader of sorts ever since high school, so it is not surprising that she was among the first to join the underground in Warsaw in 1941 and eventually became an officer. It took a number of months to convince Ida to tell her story for this book. Her hesitation was a result of her belief that she had done nothing out of the ordinary, nothing heroic, and that most young people in Warsaw during the war did the same. Eventually she changed her mind, not only because I assured her that I would record her story as "representative" of what young women were doing, but also because she became angry when, at a Polish veterans' meeting, women's participation was dismissed as non-combatant. It made her angry enough to write about her experiences for the first time and have them published in a Polish newsletter,[2] and to call me up and give me not only her own story but connections to many other women she knows, including Black Barbara. Now she says, "I'm glad you are writing this book. For years we sat by quietly and let the men talk about their war. Now it is our turn, and it's about time people took notice of what we did."

Black Barbara, Irena Komorowski, agreed right away to talk to me. "Write anything you want about me, I'll be dead by then anyway," she said, beginning the interview in laughter. Her nature hasn't changed in the forty years since the war, when, she says, she was "young, humorous and a lover of poetry." Irena is tall and still slender. Her large brown eyes are magnified by thick glasses, which are the more notable because she had "vision like a falcon" when, as a young woman, she navigated routes to safety through the battlefields of Old Town, Warsaw. Irena is charming, lucid in memory, full of details and at times very funny. She is also passionate about experiences she will never forget and about the destruction of her country. As for her own story, Irena displays a nonchalance about its publication because she, like Ida, believes it is neither unique nor especially heroic for those times. When I read to her passages from another book[3] describing her "heroism," she told me she had not read it, but laughingly added, "I suppose what he says about me is true."

Ida and Irena were affected in a different way from the others in this book by the outcome of the war: they chose not to live in Poland under Communism. They became Canadians. They still, however, spend a lot of time at the Polish Community Centre in Toronto. On the walls hang huge oil canvases of Poland's great cities before their fall in 1939. That is how they choose to remember.

Only seven days after Hitler's forces crossed the border and invaded Poland on 1 September 1939, Warsaw was under ceaseless bombardment. There was no electricity, no transportation, no radio broadcasts. Apartment buildings were levelled, and those who survived described the sensation of an earthquake as seven-storey buildings swayed under attack. Irena Kwiatkowska, a young medical student at the University of Warsaw, moved through the suburban neighbourhood she lived in, giving first aid to the injured. "Our suburb was caught between the last Polish lines of defence and the Germans," she recalls. "We lived in the basements, without light and without any communication from the outside. The bombing and artillery barrages lasted until September 27, then suddenly fell silent. We left the basement and went upstairs, out onto our balcony. Somebody said, 'Look — Warsaw is burning.' The sky above the city was pitch black, as though a hurricane was approaching.

"I walked about one kilometre to where the last Polish troops were. The soldiers were taking down the barricades. I said, 'What are you doing?' They said, 'We have surrendered.' I said, 'It would be better for us to die fighting,' and began to cry. People around us were crying too. Of course we had no idea how really inhuman it was to be.

"There was a lapse of a few days; then the Germans entered Warsaw. They looked as if they had never seen battle. They were immaculate, in perfect formation, as though it was a military parade. They were loud, laughing at our poor defeated soldiers."

Poland's fate had been sealed in only a few weeks. Her ill-equipped and outnumbered armies began haphazard retreats to the east, where they fell squarely into Russian hands. Soviet troops crossed the eastern border on 7 September, absorbing the eastern third of Poland into the USSR. The western third of the country was absorbed into Germany, and the greatest portion in the centre, containing twenty million people, fell under German occupation. Polish soldiers who had escaped capture hastily buried their weapons and uniforms in the still-soft autumn ground. Some escaped the country, aided by networks that had sprung up immediately. Thousands of men were guided across the Carpathian mountains into then non-belligerent Hungary in perilous night journeys led by local women who had skied the mountain passes since childhood. From Hungary they made their way to France and England, where they joined the Allies under the leadership of General Sikorski.

In Warsaw, what was to become the pattern of life under the Nazis began immediately. Jews were rounded up and herded into a ghetto that would eventually hold approximately half a million men, women and children. Polish gentiles who were old enough and fit for labour were deported by the thousands to work in labour and agricultural camps in Germany. Hitler detested the Poles, believing that they were an inferior race further contaminated by the presence of a large percentage of Jews. Polish fascists collaborated with the SS, spying on those who resisted, rounding up and shooting Jews who had managed to escape the ghetto. Fear and distrust were synonymous in occupied Warsaw.

As in most occupied countries, it was in the universities that the first seeds of resistance were sown. Irena: "From the beginning we spoke of resisting. I associated with a group of students at the university who were trying to organize some kind of effective work. I think that all young people, no matter how patriotic, despised the loss of freedom and were prepared to do anything to get that freedom back. I wasn't any different than the rest."

The two largest bodies of resistance in Poland were the Communists, who organized large partisan armies throughout the country, and an organization, led by former Polish officers who recognized General Sikorski's leadership, called the Home Army (AK).

Forty thousand women would eventually be official soldiers in the ranks of the Home Army, a number explained in part by the compulsory military training girls had received since 1937. Ida Kasprzak, then a voluptuous blonde of eighteen, remembers: "All high school kids participated in the training, which was like cadet school. In the summer we went to camps in the forest. We learned to camp, to shoot guns and even how to use gas masks and drill. I loved it — there was a real feeling of togetherness. I think our government had instigated this because Poland is in such a precarious position geographically, between Russia and Germany, and we didn't trust either. By the 1930s we could see Hitler building his factories, his guns, and we knew about his dreams."

Both Irena and Ida joined the Home Army. Irena, then twenty-four, was tall and long-legged with dark brown eyes and black hair. She had excellent marks in university, in part because she has a photographic memory, a major asset for the clandestine work she would do. She joined the AK after a family friend approached her. "He said they needed someone to work for AK headquarters. It was more specific than what my university group had been doing, so I thought about it and said okay."

The first contacts that led to the formation of the AK were carried out by women. Janina Karasiowna, known as Bronka, became the head of the communication department and was responsible for recruiting large numbers of women to staff it. "We had meetings in each other's houses, never at the same address twice in a row," says Irena. "We didn't know who we could trust. So we were an enigma, here, there and nowhere." The members took network names to help preserve their identity. Irena chose "Barbara," a popular children's story heroine who was an admirable fencer and a good shot. "But it was such a popular choice," she laughs, "that they had to distinguish all the Barbaras from each other. They called me Black Barbara because of my dark hair."

When the AK became "official" in 1942, the members were

asked to swear military oaths. Black Barbara: "It was a cold day, and they all arrived with crucifixes and candles. They lit the candles, and the chief started reading the oath and I had to repeat it. I still remember the last sentence, 'Victory will be the prize, and deceit will be punished by death.' So then I was really under military rule."

Black Barbara helped with organizational work and also carried messages from unit to unit throughout Warsaw. She described how the AK had begun to organize: "We split into many groups — intelligence, sabotage, administration, arms manufacturing, arms buying and arms training. Young boys who had been fifteen or so when we were invaded now were sent to the forest to train with weapons. Others learned explosives. A women's sabotage school was begun in 1942, and women also became members of the Grey Ranks, commando units specializing in diversionary actions and sabotage. In administration we planned strategy, how to take over small towns or railway stations. We printed information booklets which we stole from German arms shipments, showing how to make guns and other weapons. The AK members who could speak German translated the booklets, and we modified the instructions and built our own, simpler versions of their guns. We also communicated the information to England.

"Women were supposed to be the brains and heart of the AK. We did most of the liaison work. For that we had to memorize phone numbers and addresses and then, as quickly, forget them."

Ida joined the AK along with her entire family — her widowed mother, then in her thirties, her older sister and younger brother. She worked in an office during the day, but during her lunches, after work and on weekends, she smuggled guns. "It was common for girls to do this," she says. "There were many corrupt Germans who sold the AK weapons on the black market. Girls would be used to carry them, under their loose coats, to safe storage areas. The Germans did not pay much attention to women in the beginning. It took them about three years to realize that we were doing these things. Girls were good at this kind of work. We have better memories in general and, I think, are better at controlling our nerves. The most important thing

I learned when I was smuggling arms or messages was that if a German stopped me, I had to look him straight in the eye. I would not take my eyes off him. It made them nervous. Once when I was caught up in a routine inspection I kept staring at the soldier, which made him hurried and embarrassed. He didn't search me well, and I left with my cargo intact."

Black Barbara was in charge of liaison for her unit. This entailed knowledge of all its operations as well as the ever-changing passwords. "I would tell the couriers what the new passwords were and give them instructions on which routes were considered most safe that day. I also organized all the paperwork. Some of the full-time members who were completely underground [not known to be in the country] were paid salaries by the AK to survive. I was in charge of all these things." Black Barbara's parents knew their daughter was involved, but not all the details. They lived in fear that she would be caught. "It was awful," she says. "The Gestapo came to our building twice. The first time my mother rushed into my bedroom and said, 'Irena, Gestapo!' I sat on the bed, shaking. We lived on the third floor, so we could hear the boom, boom, boom of their boots on the stairs. They knocked on someone else's door and took a man away. My mother almost fainted after they had gone. I thought a lot about being caught. I even began to practise jumping off streetcars. In my naïveté it didn't occur to me that they took you to Auschwitz in cattle cars, not on normal trains, so knowing how to jump off trains wouldn't do me any good."

She also recalls developing the resources to deal with her nerves, and skills for the work. "I knew how to get to any address in Warsaw on foot, and Warsaw is huge. I developed perfect twenty-twenty eyesight. I had the vision of a falcon. I could see or hear up to three blocks ahead, if there was anything unusual going on in the streets. I developed strong instincts for danger.

"One drizzling ugly day in November I got into a streetcar which was half empty. I felt something was wrong, since the streetcars were usually packed. I was carrying secret messages written on cigarette paper, which I had hidden in my muff. I was preparing to get off at my stop when we heard someone yell

'Budi!' which in Polish means 'German covered truck.' German troops blocked off the whole area and were searching everyone. We had to line up in front of one of the German soldiers. I put the muff in my sleeve. My heart was pumping. I think for a couple of seconds I was one hundred percent conscious of every detail. I can still see the ground, soggy beneath my feet. Soldiers lined up pointing their machine-guns at us. And then I saw a tiny little German; he was wet and cold, and he wore metal frame glasses. He looked tired. I stood in his line. When my turn came I didn't have a drop of saliva in my mouth. I couldn't speak. I handed him my purse which had my nurse's papers in it. My knees were bending as though they were made of gum. He told me to go. I almost sagged. When I started walking away a lady approached me and asked what was happening. I said, 'Please, leave me alone. I don't have the strength to talk.'"

Ida also described the tension they lived with: "I was constantly afraid. I left home every morning and did not know if I would return. There was a fifty-fifty chance. There were spot-checks at random throughout Warsaw, and anyone who didn't have proper papers would be sent to Treblinka or Auschwitz or Madanek. If they discovered you worked in the underground you'd be taken to Gestapo headquarters, tortured and shot."

The AK targeted German officers for assassination. The assassination squads were often Girl Guides, who would steal into the officers' houses and plant explosives or shoot them in bed. These girls carried cyanide capsules because chances of their capture were high. Girl Guides and Boy Scouts also served as couriers for the AK.

Inside the Warsaw ghetto thousands of people were dying of starvation and disease. The Home Army kept couriers going in and out with messages and small quantities of food. Small numbers of men, women and children were smuggled out via the underground and hidden in cellars, then helped through escape routes to the forests. The AK courier girl who went to the ghetto regularly was killed on one of these missions. Ida was sitting with her commander when the news was brought in.

"The officer came and looked directly at me as he told us about her death. In an extremely loud voice he said that an urgent message had to be delivered, and proceeded to repeat it word for word. I pretended I was not listening. I knew he was saying it for my benefit, even though he wouldn't come right out and ask me to go. I didn't even look at him. I was thinking, 'I'm a young girl, only eighteen. I am not ready to die. I refuse to die for anybody.' He repeated the message a third time, and of course I knew I had to go. I had no choice. So I said, 'Okay, start from the beginning,' and I repeated it after him. He was still in shock from the death of the other girl, and he forgot to tell me what route she had taken or how she got killed.

"I went via the underground. By this time people were knocking out the walls of their basements in the row houses. One could traverse whole city blocks, literally underground. I arrived inside the ghetto and delivered the message. As I began to leave, a woman approached me with her two small children. She asked me if I could smuggle them out. I said, 'Yes, I will try.' But the moment I took their hands and she started to walk away, they started screaming and kicking. They were completely upset, they didn't want to go with me. They kept trying to run back to their mother. An elderly man came in and told her it was not a good idea to let them go. They were so young and scared that they might scream the whole way back. People would realize they weren't my children, and it would endanger all of us. So she took her children and walked away. That was the only time I visited the ghetto."

By July 1942 the Germans had begun clearing out the remaining four hundred thousand people in the ghetto, transporting thousands a day to Treblinka, Auschwitz and Chelmno. Rumours about the extermination camps were confirmed when trainloads of human hair revealed traces of hydrogen cyanide. In January 1943 those remaining in the ghetto, numbering approximately fourteen thousand people, resisted. They met with gunfire the Germans who came to deport them. The ghetto uprising lasted until May, and when it ended only one hundred men and women managed to escape through the sewers. The others were killed or captured. The Germans bombed out the remaining shells of buildings, and all that remained was a moon-like landscape of craters and rubble. The Warsaw Ghetto Uprising is a story for another book. Many of the fighters were women. A handful of survivors live together today in a kibbutz in Israel. Part of the sorrow of the story is the lack of help from the outside. The AK offered some assistance, but it was limited. There were AK members who joined the fight and were killed, but their numbers were few.

Ida says: "I know sometimes Jewish people have something against us, but we couldn't do anything. It was not the time to rise up. If we had, we knew we would fail. It was hopeless." She adds, "Our group bought guns from German soldiers and got them to the ghetto. That's all we could do. It was terrible. We could hear the shooting in the distance."

German hostility was now focused entirely on the Polish underground. Black Barbara remembers that the atrocities grew to have less and less impact on her: "I felt as though ice was forming over me, slowly as ice does. The first layer is thin, and then, if it is cold enough, it grows thicker and thicker. One day my father came home white as paper and began to vomit. We found out that he had been on a streetcar and passed by a street where some boys in the underground had been printing illegal papers. The Germans had caught them and had hung all fifteen boys from a balcony. My mother started to cry when she heard this story. I didn't say anything. Then my mother got angry with me because I wasn't crying too and because I didn't want to listen. At twenty-five years old I couldn't let myself become too emotionally involved because I didn't know whether it would be me in the same situation the next day."

On Ida's twenty-first birthday, 17 January 1944, she received a memorable gift. Arriving at her office in the morning, she was excited to see a large bouquet of flowers, amid which was perched a Je Reviens perfume box. She opened the card and read the words of her admirer, marvelling at his ability to get, not only her favourite perfume, but any perfume at all in 1944 Warsaw. She noticed that the box was a little heavy and opened it. "There, lying in the beautiful white-satin-lined box, was a revolver. I almost fainted. Thank God nobody was standing near me."

Ida carried the weapon in her handbag with mixed feelings of security and fright. Though an expert shot, she wondered if she would ever be capable of using the gun. "Women have babies, and I thought maybe I wouldn't be able to pull the trigger." On one occasion she discovered a new instinct in herself. "Some girls from my unit and I were going through the underground to get to a meeting. By then the underground went miles and miles, and we even put street signs down there. I had the gun in my purse. Suddenly we saw two men with big heavy sticks in their hands coming towards us. They were not Germans but drunken Poles from another group who were against the AK. They started to yell at us, and there was blood on their sticks. We immediately thought they were going to rape us. I pulled my gun out of my bag and said, 'Okay, the first one who moves is dead. I'll shoot you.' I stared them right in the eye and said, 'Don't doubt me, I will shoot to kill.' They stopped. I saw a doorway and ordered them to go into it. Once they were in it I shouted to another girl to lock it. Then I looked at my hand which was holding the gun. It was shaking, and I saw that in my nervousness I had forgotten to cock it. Lucky they had been too drunk to notice.

"We got to the meeting and I told my commander about the incident, saying we had marked the door with an *X*, and someone had better go get them out later. And do you know what my commander said? He was furious that I owned a gun and hadn't handed it in to them, since the AK was desperately short of weapons. They didn't think a woman should have a gun when lots of the men didn't. But I refused to give it up. I said, 'It's mine. It was a present, and I shall keep it,' which I did."

By early 1944 the AK was ready to go on the offensive. Encouraged by German defeats in the Soviet Union they instigated a general intensification of armed actions throughout the country, which they code-named Operation Burza, or Tempest. Sabotage groups stepped up their nocturnal missions and partisan armies launched isolated and sporadic attacks, but the operation in general was poorly co-ordinated. Ida's young brother, then only seventeen years old, was killed during Tempest when German soldiers surrounded his AK partisan camp in the forest. Ida had been very close to him and was very bitter after his death. From that moment on she believed that he watched over her and protected her.

As Soviet armies moved eastward into Poland, the AK considered launching an uprising in Warsaw to secure the capital before the Soviets arrived.

The AK hoped to receive the Russians into their city as guests rather than liberators. The plan was neither fully encouraged nor outwardly objected to by the Polish government in London. General Sikorski's refusal to acknowledge any Soviet right to lands they had occupied in September 1939 was creating a serious rift with Moscow that worried and even annoyed the Poles' British allies. Stalin was never flexible regarding Poland, and relations could hardly have been worse for co-operation on the battlefield. Furthermore, the AK motive for an early rising in Warsaw was not misread in Moscow. And finally, the AK blundered in its estimates and timing. The rising was planned to coincide with the arrival of Soviet armies in Praga, the eastern suburb of the city, which lay across the Vistula River from the downtown core. On 1 August 1944, when the order finally came to rise up in Warsaw, the Soviet troops were still many miles from Praga, though their gun-fire was audible in the distance. There they stopped, and for reasons still debated by historians refused to advance in aid of the uprising. Warsaw, without help from the east, and impossible to adequately supply from the west, rose up in a battle that was doomed from the start.

Inside the city the stirrings that accompany all rumours of battle were particularly stimulating to the people who had lived under German rule for five long and terrifying years. The AK had mustered up as many light guns and as much ammunition as they could, but it was enough to supply only an estimated one out of every ten of their forty thousand soldiers. The odds were clearly against them, but the people of Warsaw were, in the hot days of summer 1944, ready for a fight.

Black Barbara, working at the headquarters where AK chief General Bor-Komorowski and his advisers gathered, was concerned with her own role in the coming fight. She asked her chief, "'When the uprising begins, what will I do? Do you expect me to push papers from one end of my desk to the other?' He smiled at me and said, 'Don't worry. When the uprising begins you will be right in the middle of it.'"

In the funny way memory lingers on the small things in the midst of the momentous, Ida remembers being upset the day the uprising began because it ruined her plans for a date. Black Barbara was given a day's forewarning and was on her way home to get some things when the first shots rang out. "There were no streetcars working at the time. The Germans were in disarray with the approach of the Russians. They were riding through the city in carts pulled by cows and horses, and everything was a bloody mess. We were on the bridge when we heard the first shots. We spent the night in a basement nearby, knowing we could never make it home. At six o'clock in the morning I got up and decided I must make it back to my chief or he would chop my bloody head off. Already Warsaw was divided into Polish- and German-held sections. The AK had seized the city hall, the post office and quite a few large buildings and blocks of the city. The Germans, caught off guard, didn't even begin using their tanks until the next day. There were parts of the city where I saw the Polish flag flying and heard our national anthem playing. It was indescribable, what I felt, how my imagination was working. I was in a hurry to get back to my unit. I didn't want them to think I was a coward or that I had deserted. I made my way across the city by trial and error. I would come to a corner and stand under the archway and look to see if I could spot any German sharpshooters in the building windows above. Then I would rush out and make it to the next archway.

"Under each archway there were crowds of people hovering. Near one of Warsaw's oldest, biggest hotels, I stopped to figure out what to do. There were four or five dead people lying in pools of blood. A sharpshooter was firing down on the pavement. People were very upset, all talking at once, and there was bombing and shelling going on. A young man, a waiter from the hotel, was extremely agitated, telling everyone the whole story of the war, how it felt and what was happening. I looked at him and suddenly I noticed that his genitals were hanging outside his trousers, quite visible against his black trousers, because in his hurry to pull them on and get out to the street he didn't realize. . . . I looked at this, and in the middle of this whole tragedy, I started laughing to myself. I've never seen anything so funny in my life. Of course I couldn't tell the poor guy; I didn't want to embarrass him. I hoped somebody else would.

"The journey I was making would normally take about three-quarters of an hour. It took me from six o'clock in the morning to two o'clock in the afternoon. When I got to the headquarters somebody came and opened the door and said, 'Barbara, the chief is bloody mad at you.' I said, 'To hell with everything. I am surprised I am still alive. He better be happy to see me.' Of course he was when he heard the whole story.

"I was so exhausted I lay down on a pile of coal and fell asleep. My first mission was to go to another unit to deliver a message. In twenty-four hours everything had changed drastically. The Germans were trying to divide the city, and we were ordered to take the northern part under our command. That was the part of Warsaw that had housed the Jewish ghetto and also Old Town, a medieval section of the city with rows of stone houses, winding streets and piazzas. Going through the former ghetto was just like walking on the moon, there was nothing left but rubble, and there were Germans shooting from the church steeple overlooking the area. Crossing through was perilous since there was no cover anywhere."

On 7 August, only a week after the uprising began, the entire AK headquarters moved to Old Town where Black Barbara was stationed. She ran messages through the gamut of barricaded streets and shoot-outs amid the rubble. No Soviet bombers came, but the Luftwaffe did, and began systematically to raze the area. Germans drove their tanks, civilians strapped on the outside, through AK barricades. The AK made the painful decision to shoot anyway.

The people of Warsaw remained in their cellars, eating what few food supplies they had stored, sharing what they had with soldiers of the AK. Organized kitchens were set up for the army, run by civilians who volunteered. Girl Guides and Boy Scouts made endless courier missions, by now using the city's sewers, as the street were impassable. Casualties were evacuated, through the underground maze of knocked-out cellars, to makeshift hospitals.

Ida was an officer in charge of a group of women medics. She and her staff were responsible for carrying the wounded to safety and applying first aid. "They didn't give us guns, because they would not have helped us. Only those behind the barricades were given weapons. Because there weren't enough to go around, they rarely offered the girls any. If it happened that an AK soldier was killed, the girl next to him would grab his gun and start shooting.

"Out in the streets we were open targets, dodging bullets, with no time to shoot anyone. The Germans had a big gun they called Big Bertha. They placed it on the railway bridge aiming down at us in the streets below. A lot of people in the area were wounded. We were sent out there to bring them in. People were screaming and moaning. My girlfriend and I spotted a man severely wounded in the head. She grabbed his legs and I grabbed his arms and we ran through the streets. He was a huge man, he must have weighed two hundred pounds. We were running, and I saw a piece of shrapnel fly into her leg. Blood was running down, and I wanted to scream at her to put him down. Just at that moment I saw a piece of metal coming straight towards my stomach. There was a gap of about six inches between his head and my stomach, and it was coming straight towards us. I thought, my God, what can I do? If I drop him he will die as soon as his head hits the pavement. I was petrified. All this occurred in split seconds, of course. Then, just as the shrapnel came within an inch of my stomach it fell down, between me and the wounded man's head. It was as though someone had punched it from above. I screamed 'Dunca, stop running, put him down gently.' I ran back and picked up the shrapnel. It was still hot. I carried it with me through the uprising, and always felt as though I was not alone, that somebody was helping me. I still have it today."

Ida's unit set up a hospital in an old shoe store. The chief surgeon was a veterinarian whose specialty was horses. The most seriously wounded patients would be taken to the main hospital, but the veterinarian did operate on those he could. "He treated the wounded like animals. He was really very good as far as operations went, but he wouldn't look into their eyes, he acted as though they were horses. I mean, he loved horses. But that is how short of trained medical people we were." There were few supplies. For anaesthetic, a liqueur cordial was used. "It was all-purpose. We used it to clean wounds, for drinking and for pain." There were wounded among her first aid women. One, a close friend, was blinded in the first days of the uprising.

Ida was in charge of a unit of thirty women. "I think I was chosen because I always had been a leader in a way, always volunteering for things, even back in high school." They were responsible for carrying the wounded from the street battles, assisting in the hospital and also carrying messages and delivering guns.

"We slept in the cellars of the houses, a few of us all together. Parts of the city were completely cut off from each other and it was now impossible to contact people in other parts. The worst thing for most of us was not knowing the fate of our families."

Food shortages in the city grew acute. "A group of about twenty prostitutes organized and began to help us. They went to an old brewery nearby, got heavy sacks of grain and delivered these all over our section of Warsaw."

Ida describes German planes flying "so low that we shot some down with rifles." The city was on fire, rocked by explosions day and night. The underground through cellars was obstructed in many places by damage from the bombing. The first aid women travelled through the portions that remained, and sometimes up through the top floors of buildings and back down again. It took hours to navigate a distance of perhaps half a mile or even a few blocks.

By mid-August the key German target was Old Town, which they bombed day and night. Black Barbara had become an expert at finding routes through the rubble and burnt-out shells of buildings, "which were the safest, because they wouldn't bomb ruins. The barricades at each street corner were made from rubble, old furniture — anything. Everywhere was blood, blood and charred blackness. If you were blond, your hair was black, and if you were dark, it was streaked with white. Everything was pulverized. The artillery noise was terrible."

Each day General Bor-Komorowski and his advisers planned their strategy. Barbara, present at many of these sessions, remembers thinking that "even Hercules couldn't win against so many enemy forces." Old Town was almost surrounded by German forces. An artillery unit was operating to the north, and German panzers came down the railway tracks to the west, bringing in reinforcements. In the fierce fighting for the medieval streets, the AK took prisoners. As civilians hid in what was left of their cellars, German POWs were put to work clearing debris and helping to build barricades. Barbara, who had sworn a personal oath to kill at least five Germans, passed by the POWs at work. "I was smoking at the time, and one of the Gerrys looked up at me and said, 'Fraulein, cigarette?' I looked around to make sure no one was watching, and then threw him the whole pack. At that moment I saw him as a human being. I think women are soft-hearted. I was so ashamed of myself. I had wanted to kill five Germans, but instead gave them cigarettes. I felt for a moment that they were my brothers in misery, even after five bloody years of occupation."

Supplies were by now being brought into the beleaguered units in Old Town via the sewers. Teenagers, mostly Girl Guides and Boy Scouts, navigated the hellish labyrinths that were still in use. Many never found their way out, drowning in sewage in the pitch blackness. The ones who made it became the lifeline for the AK unit that clung to the territory, which consisted of less than half a square mile. "Those young boys and girls were responsible for getting us ammunition," says Barbara. "They came up, covered in slime, handing us grenades and bullets for our machine-guns." German 600 mm artillery guns crushed the ancient houses and left craters in the cobble-stone streets. In the choking, thick, blackened air, Barbara says, "I could see only a foot away." The sounds of destruction mingled with the cries of civilians buried beneath their collapsed homes. Those who survived took refuge in the ruins.

Barbara remembers one mission vividly. She went with her chief to deliver a message, and they had to cross a large piazza: "Around the piazza there were tiny little houses all smouldering or still on fire, and behind the barricade about fifteen people were gathered. When we got there I told my chief he couldn't go with me. I told him there was no chance he would make it, and he would get me killed as well. I put the message and maps in my bra, along with about twenty bank notes I had to deliver. I was wearing a little gold medallion of St Anthony around my neck. I counted down from ten, got to zero, took a deep breath and jumped across the barricade. I was running. . . . I could hear machine-guns going off, and I heard a rip as though somebody had torn material from my coat. I kept running. . . . The machine-gun rat-tat-tat didn't stop. . . . I saw something like a door ahead, blurred in front of me, and I thought it would make a shelter, but when I got to it, it was boarded up. I was exhausted, because it was a big piazza. I turned and looked towards my chief, and he signalled me to drop down. I threw myself to the pavement, and as I did I saw a man lying nearby, shot through the head, and all his brains were lying on the pavement. I thought the bloody Gerry would do the same to me. I began to run and drop, jump and drop, the way soldiers do, from one heap of rubble to the next. I thought, to hell with you, Gerry, you won't get me easily. I realized the only thing to do was to get back to where I had come from. I was looking around, trying to figure out where the firing was coming from, and far away, in the fifth floor of a building across the street, I saw a machine-gun in the window. I had very long black hair and I knew it made me more of a target, so I tried to cover my hair with my coat and keep my face down close to the pavement. I crawled a bit, then lay flat. He couldn't get me because I was at a difficult angle and quite far from him, but he kept shooting all the time. He would hit the sidewalk near me, or a piece of glass, so I had things shattering and flying up around me. I crawled a little farther to where there was a broken piece of curb, pavement, about the size of my head. I crawled with it, pulling it along in front of my head. He shot some more and I thought, to hell with you, I'm dead anyway, and I just lay there. I didn't move for about five minutes. He stopped shooting. When I moved again he shot again. It went on like this for what felt like an eternity. It was probably about twenty minutes in all. I could see my chief waving his hat at me. The others behind the barricade had disappeared. Ahead I could see a telephone box, and I thought if I could make it that far. . . . Then there was something on my left I was counting on. It might have been a street lamp or a tree, I don't remember, but I thought if I could get to that spot I could jump up and get away.

"Just as I got there, I heard someone yell, 'Halt! Hands up!' I stood up, put my hands up and saw three German soldiers pointing their guns at me. I began walking slowly towards the Germans. I had to pass the barricade. I kept my hands up and kept walking right towards the corner where the barricade was. . . . The Germans must have thought there was another German soldier there, I think. I got to the barricade and threw myself over. . . . So, with my hands up, I walked back to safety. It was amazing. The Germans couldn't follow me or they would have been shot by our sharpshooters.

"An old civilian woman rushed up to me with a bucket of water; she was sure I must be shot through several times. My colonel asked if I was wounded. I said, 'I don't know.' All I wanted was a drink. My tongue felt swollen and I couldn't move it.

"I looked under my coat and dress. The lace of my bra had a burned hole right through it — a very small hole, like a cigarette burn — but I had been saved when the bullet hit my medallion and all the papers."

Across the city, in an area near the Vistula bridge, Ida's unit was engaged in fighting for buildings floor by floor. One side of the street was controlled by the AK, the other by the Germans. In the humid heat bodies lay rotting on the pavement. No one could clear them away because of the fighting, and the stench of death became almost unbearable. The International Red Cross workers who normally removed the dead were unable to get in to the street. Two bodies lay there for almost a week before Ida's colonel called a meeting. The smell was awful, and yet he could not risk sending down his own men to what meant certain death. Ida suggested that four strong women could probably do the job.

"He looked at me as though I were crazy at first. Then I saw his mind beginning to work. He got a sparkle in his eyes, you know, like a little boy. He looked at me and said, 'Can you find four heavy women?' I wasn't sure that I could, because they would have to be volunteers. We all knew that there was only about a ten percent chance of coming back, so I just lowered my head. Then I said, 'I volunteer, and I'll try to find three others.'

"The colonel gave a long speech about how dangerous it was, then turned the meeting over to me. My best friend, who later lost her eye, volunteered along with two other girls. We put on white overcoats, attached red crosses to the arms and painted one on the sides and back. We took shovels and put our ID cards in our pockets in case we were caught. We took a back route, behind the building and across the railway bridge viaduct. We got to the German building.

"We spoke to the German commander, told him we were Red Cross and we wanted to clear the bodies. He said, 'You are really Red Cross? Those bandits across the street will kill you if you go into the street.' He asked for our Red Cross cards, and we pretended we didn't know what he meant, and gave him our Polish ID. He started laughing, gave us back our cards and said, 'Okay, but one of the soldiers will go with you and guard you.'

"We walked into the street and felt about two thousand pairs of eyes staring down at us from the huge buildings along the road. Everyone held their fire. Maybe because there were so many people watching, my hands were so stiff I could hardly move them. We dug the graves. The bodies were barely recognizable. I pulled one by the shoes, but the shoe stayed in my hand. We had to lift their bodies with shovels, piece by piece into the graves, because we couldn't touch them. Then we were marched back to the German building.

"The officer then said, 'There is another civilian body near the river. If you are really Red Cross, go bury him too.' We had no choice. Just as we were finishing two German soldiers approached from a different direction. They spoke to our guard. They asked, 'You are Red Cross? Of what nationality?' We said, 'Polish,' and they said, 'We know you are Polish girls, not Red Cross. Only Polish girls could be so stupid.' Then they ordered us to walk with them as their shields for about two miles to another section of the city.

"Our AK soldiers were furious, because they couldn't shoot them. The Germans pointed their guns at us and our soldiers did not fire. I prayed that nobody would be nervous and pull a trigger, because one shot would have started a whole war again."

Ida and the three other women walked with the soldiers to the statue of the Mermaid, at that time Warsaw's most famous monument, which overlooks the water. That was as far as the soldiers needed to go. "For a terrifying moment, we thought they would now kill us, but they walked away." The women walked back to their unit.

"When we got back everyone gathered around, hugging and kissing us. They made heroes of us, and I don't remember hearing this, but later I was told that my commanding officer said, 'How can victory fail us when we have women like this?'"

By 26 August General Bor-Komorowski had moved AK headquarters out of Old Town back to downtown Warsaw. Three days later Old Town fell into German hands. On the last night Barbara's commander received orders to leave. It was an emotional moment. At first he refused, but orders were orders, even in the makeshift armies of the resistance. Barbara began to cry. "I was petrified of the sewers. I had a fever that day, and my period. I was afraid I would get blood poisoning or a disease down there."

They descended into the darkness of the nineteenth-century sewers, which were less than five feet high in some places, only a yard high in others. They walked through knee-high waste, holding on to a string. In the blackness people cracked their heads against the walls and cut their feet on jagged debris. It was a nine-hour journey: "We had to stop just before every manhole and proceed very quietly so that any Germans stationed above would not hear us and throw down a grenade. If someone pulled on the string once, it meant stop; twice meant to continue again.

"We finally emerged. It was a dark, dark night, and the streets were quiet because they were bombing another part of the city. We were exhausted. I turned to one of my companions and said, 'Jacques, what will become of us now?' He said, 'Shit. I don't care anymore.' At that time men were not supposed to swear in front of women. I was so upset and exhausted from the sewers that I smacked him across the face. I cried, I was very mad. Everybody started asking what happened. Someone said, 'Irena smacked Jacques in the face.' Someone else said, 'Good, he deserves it, he is always so rude.' About two hours later they took us to a school to sleep, where there was warm water and a floor to sleep on. Jacques came to me and said, ‘I’m sorry, Barbara. Don’t be mad, I was so tired.’ He brought me some alcohol, about half a glass of vodka. I gulped it down, lay down on the paper and slept like a baby.”

After over a month in the severest fighting in the city, Barbara was astounded to find people still sleeping upstairs in their apartments. She walked to General Bor-Komorowski’s headquarters the next morning without dodging bullets. “When I got there, there was a boy of about nineteen holding a gun, guarding the entrance. I said, ‘I have to go inside to see the general.’ He asked, ‘Do you have your pass?’ I said, ‘No, I have just escaped from Old Town. I have to report to General Bor immediately.’ He said, ‘Please go away. I am a soldier. I am busy.’ I shouted, ‘What? You are a soldier? You are a clerk — I am a soldier!’ Later I was very sorry that I had said that because he had such a pained expression on his face. I really hurt his feelings. What I said was true, but I shouldn’t have said it. Men have such fragile egos about these things.”

By 4 September the Germans had begun their assault on the city centre. A 600 mm shell burst through General Bor’s headquarters, but miraculously no one was injured. Headquarters moved again, this time to within three hundred metres of German lines. In the weeks that followed, the problem of the wounded was compounded by those who were starving and dying of disease. Thousands of homeless refugees descended upon the AK, which had neither the resources nor the time to deal with them.

Working furiously in the shoe store hospital, Ida remembers that there was no more room to put the wounded and no more medical supplies. In the streets AK soldiers were running out of ammunition. Working with little sleep and less food in the desperate surroundings, she remembers feeling hopeless. “We wondered if we could go on. Lifting the heavy bodies, there were points during the attacks when we would just sit down and couldn’t move. I couldn’t even eat what food there was, I was too exhausted.”

Food was no longer to be found in the city. On 10 September, the Soviets finally made air drops, but without parachutes. The long-awaited delivery arrived crushed, and much of it fell into German sections. Only in mid-September would Stalin finally permit U.S. pilots to use Soviet air bases to help supply Warsaw. The one American drop that followed was too little and too late. Inside the city people killed their cats and dogs for food, as the city fell, block by block, back into German hands.

There were also innumerable and unknown instances of real heroism. Particularly notable were the fighting units of Girl Guides and Boy Scouts who had fought battles in Old Town. Some civilians contributed to the fight generously: at a time when a piece of bread could mean life sustained for a few more days, some hoarded and fought for it while others gave it away.

Ida: "We were out looking for wounded, and passed through the courtyard of an old house. There were two of us. An elderly lady emerged from inside the house where she was living completely alone. She must have been about eighty. She said, 'I saw you girls earlier, passing through here, and I've been waiting for you to come back.' We looked at each other. She went back into the house, came out with two cooked chicken drumsticks and gave us one each. Both of us started to cry. She said, 'Why are you crying?' We told her we hadn't eaten in such a long time. Chicken was an unheard-of luxury. We sat on the steps and afterwards she blessed us and kissed us both on the forehead. We wanted to take her back to the hospital. She refused, saying she had been born in the house and wanted to die there. To this day I can still see her face, standing there with those drumsticks. I never cried when I was wounded. This was the first time I had broken down during the uprising. Kindness always gets me down."

Barbara remembers seeing others crying, including many of the men. Things were becoming more desperate, and the women began to reach out more to comfort the men who were beginning to fall apart. Her own colonel would later say of her, "I must mention the girl who was my own messenger during the conspiracy and fighting for Old Town. I am referring to Black Barbara. . . . In fact, she was not only a messenger but also an adjutant in the full meaning of the word. She was tremendously courageous, ready to sacrifice herself."[4]

On 20 September one of the last pitiful messages from Warsaw AK headquarters reached London. It said, "Warsaw is in ruins. Street after street is being systematically razed to the ground by artillery and aircraft. Nobody uses the streets. All communications take place by means of interconnected cellars. For a long time there has been no light. The water mains are damaged. Many wells have been sunk, but there is still not enough water. The shortage of food is badly felt."

On 26 September the message was brief: "The population is starving. People are eating dogs." A British airman stranded in the city reported that the worst was the smell of rotting bodies that permeated the centre of the city. He said, "The soldiers fighting to defend their battered barricades are an awful sight. They are mostly dirty, hungry and in rags. There are few who have not received some sort of wounds."

On 27 September, after two months of fighting, General Bor called together his staff and announced he would end the tragedy. There were no more bullets, no more food to feed his soldiers. Only two small sections of the city remained in AK hands. In the final tally close to two hundred thousand civilians were reported dead, and others were dying of disease and starvation. Of his own forces, a third had been killed.

The order to lay down their arms filtered through the city to where the last pockets of fighting continued. Surrender was formally declared on 2 October. Ida remembers, "Everyone was crying. We were so upset. We had fought for sixty-three days, and it was all for nothing. We were so furious and felt so betrayed by the Russians. It is hard to describe. We felt as though our life was over. It no longer mattered what happened, because everything we had fought for was gone."

In the round-ups that followed, all AK soldiers were taken prisoner. Ida's mother, who had fought in another part of the city, had been taken to Ravensbruck weeks before when her section of Warsaw was recaptured. Her sister was taken into custody as well. Barbara joined a line of about a thousand women prisoners who were deported to Germany. Ida and her friends discussed an escape, but she decided to stay with her blinded friend. " I naïvely believed the Germans would allow her some medical attention," she says, "but she was forced into captivity with us."

"We were the last regiment to be deported. We left on October 5. We walked about twenty-two kilometres to the railway station that was still operating. I was despondent. None of this mattered. I was completely spent, just living on nerves for so long that the moment the guns stopped I felt something was wrong with my ears."

Thousands of Home Army women were taken to the camps at Bergen-Belsen just south of Hamburg, Germany. Originally a POW camp, it had become by 1944 a collective depository for concentration camp prisoners from all over Europe. There were no gas chambers at Bergen-Belsen, but thousands of people starved to death or died of disease. When the women from Warsaw arrived there were fifteen thousand prisoners in total in the various sections of the camp. By mid-April and liberation, sixty thousand people were crammed into the quarters, among them ten thousand unburied corpses. With the approach of the Allies administration within the camp virtually ceased. Guards remained, but there was little or no food distribution.

Ida's first memory on the road to Bergen-Belsen was of young German children, seven or eight years old, who stood by the roadside and threw stones at the new inmates. The years of underground work, the tensions and the tragedies she had witnessed and suffered finally came to a frightening head within the barracks, where two hundred of the four thousand Polish women lived cramped, crawling with lice and dying of disease. Ida, an officer in the AK, was selected to be commanding officer of her barracks. She ordered the women to make the best attempt at cleaning themselves and their bunks, "a small fight against the lice and disease." But one girl consistently refused. "I tried to explain to her why we had to do it. We couldn't give in to the filth. She refused. I got so mad one day that I grabbed her and began shaking her, banging her head against the wooden bunk again and again. I was completely crazy. My friends pulled me off her and screamed at me, 'What are you doing, you are killing her!' They said if we started killing each other the Germans would not have to do it themselves. I walked out of the barracks, around and around for about two hours. Then I came back in and apologized to the girl. She said, 'Ida, I'll never refuse to help again.' I was so upset and scared. I really didn't realize I had that in me, the ability to kill someone with my bare hands. It was the whole situation. Everything. We were starving. The camp was death and hunger, and I just couldn't cope anymore."

Black Barbara and Ida met for the first time at the camps. Black Barbara still remembers those seven months clearly: "I remember the awful hopelessness represented by the barbed wire. There was a kitchen where the poor Soviet soldiers had to work. Then there was a road down which Ida's barracks were. . . .

"I thought we would die there, but not from starvation. I thought we'd be killed by the planes which were dog-fighting above us. One day I was standing outside the door of my barracks looking out when I heard someone scream, 'Black Barbara, go back inside, quickly.' I hid in the doorway and watched as American planes zoomed down and began machine-gunning the compound. A beautiful seventeen-year-old girl was killed immediately. Another girl lost a leg, another an arm and one had two shots through her lungs. There were sixteen girls killed or wounded." It was the final, tragic irony.

Both Irena and Ida were sent to England after the camp was liberated by General Patton's army in April. They both continued to serve as officers and were posted to Germany. While there Ida learned that her mother had survived Ravensbruck, when she accidentally discovered her mother's name on repatriation lists she was typing for the army.

Working in defeated Germany aroused many emotions. For Irena it was cleansing. "I saw how devastated the country was, and how they were hungry and suffering. I stopped warmongering. I felt justice was achieved to a certain extent."

Ida: "People wanted to forget. For me it was difficult. At first I tried to avoid any contact with Germans whatsoever. Then one day I attended a Polish mass. The priest told us, 'Every time you see a German, just repeat the Lord's Prayer to yourself, especially the line, "Forgive them their trespasses against you" because God says you have to forgive.'"

Warsaw was "liberated" in April 1945 by Soviet troops and Polish regiments trained in the USSR. They entered the bombed-out, mangled remains of what had been one of Europe's most glorious cities. The rubble stood six feet high in places. It would take an extraordinary effort to clear it all away and begin to rebuild. Sections like Old Town were lost forever. Underneath the twisted metal and blasted rocks lay thousands of the city's unknown dead. In all, of the approximately 1 million people living in Warsaw before the uprising, 250,000 were killed or missing. Another 350,000 were forcibly removed by the Germans during the sixty-seven-day battle. One hundred and twenty thousand remained in Praga suburb under Soviet control, and 280,000 remained living amid the ruins. It has been estimated that eighty-five percent of the buildings were damaged. The last of the annihilation came in the pitched battle between Soviet and German armies. The city that had been the first to fall in the world war was literally at war until its final hours. A Soviet woman pilot, flying over Warsaw just before the end, would tell the author, "The fires of the city reached up and seemed almost to melt the wings of my plane. I could feel the heat on my face. Below me was utter devastation."

Neither Irena nor Ida returned to live in Poland, both preferring life as immigrants to life under Communism. They both suffered considerable post-traumatic stress, which manifested itself in psychological and physical symptoms. Irena still has nightmares, though they have become less frequent as years have passed. A smell or certain kinds of lighting trigger frighteningly real images from her photographic memory.

When Ida had a physical examination at age twenty-seven, six years after the war ended, the doctor told her she had the stomach of a ninety-year-old woman. Ulcers and nightmares have continued to plague her for the past four decades because she says, "I always held in all the emotion and fear." She has stayed in touch with her old friends in Toronto and attended reunions in London, England. Once she went back to Warsaw for the anniversary of the uprising, held each year on the first day of August. "In Polish there is a word like 'colleague' that means someone who has shared the same experience with you. It is not exactly the same as 'friend.' It is something even deeper, because only you can understand each other fully. We still feel that way about each other. I don't let go of that."

THE EASTERN FRONT

Soviet Union, 1939-45

I think women's presence at the front had a dramatic effect on the soldiers. They tried to look presentable, they managed to shave themselves and not to swear too heavily. . . . It brought out their gentle side, when normally they would have been brutal.

Woman soldier, *USSR*

One of the sailors walked up to me and handed me a baby's pacifier. He said, "When we go into battle, we won't have time to baby you, so take this." I said, "We'll see who is going to take care of whom."

Katyusha Mikhaylova,
Marine, USSR

There was an enormous strain on the nerves. I would see images of burning planes crashing with my girlfriends in them whenever I shut my eyes.

Nadya Popova,
Bomber pilot, USSR

Every 9 May the small park in front of the Bolshoi Theatre in Moscow is packed with people. Crowds gather with radios, cameras and flowers to celebrate the national holiday. In Gorky Park teenagers line up for amusement rides while disco music blares from enormous radios. But most apparent in the crowds are the thousands of older men and women, now stout and weathered, who wear some kind of military dress, heavy with medals. They walk holding onto each other's arms, some still together after a front-line romance, others just old friends.

Children hand them flowers; younger adults bring them photographs taken at the festivities the year before. This is the celebration of the end of World War II, which the Soviet people call the Great Patriotic War. Despite the passing decades it is still one of the biggest holidays of the year. Clearly, these are not forgotten veterans of a forgotten war.

The women stand out clearly in the crowds. They look classically grandmotherly, in print dresses, smoothing their hair and clutching their handbags. The medals pinned to their chests seem odd, out of place. It is difficult to imagine these plump, smiling matrons as front-line soldiers. Yet nearly a million of them fought in the armed forces and partisan armies.

It was a war on an unprecedented scale. Vast armies launched tank battles that seared the earth and left smouldering, broken landscapes covered with oil, blood and twisted metal. A third of the country was occupied, and savage partisan warfare raged through the forests where technology was supplanted by primitive battles on horseback and sometimes hand-to-hand fighting. A tenth of the population of the country, or twenty million people, were killed. Every able person was needed to fight the technologically superior enemy, including women, who served in all roles and fought on every front.

According to expert Dr K. Jean Cottam, who has devoted many years to the study of Soviet women in World War II:

> More than 800,000 Soviet women and young girls served at the front — including 27,000 partisans — constituting about 8% of Soviet military personnel at the end of 1943. They mastered almost all military specialties; for instance, they became members of aircrews, tank crews, and gun detachments. In addition, many were front-line medical personnel, some of whom fought and all of whom were usually endangered.[1]

I travelled through the country and interviewed women who had served in many different capacities, from partisan to bomber pilot. Most of them wanted to talk not about themselves but about the war and its effect on the country, on civilians — to give me some sense of the scale of destruction and loss of life. When I asked about their personal experiences they showed a modesty that is frustrating to an interviewer, yet endearing. Simply, they fought out of necessity.

In Russian the word for "bravery" is masculine; it is difficult to use in connection with women. Although Russian history is full of martial women, such as those who fought in the Napoleonic Wars and in the Revolution, each woman I interviewed said that women do not belong in combat — that it is physically too difficult and that only in case of national emergency should women again take up arms. The Soviet army today reflects this opinion: there are said to be fewer than ten thousand women in its ranks, and they are designated non-combatant.

Of the thirty women I interviewed it was difficult to choose only a few to feature. They all had remarkable stories. I selected a bomber pilot, a marine and an army nurse to represent the three traditional forces, but each woman I talked to helped provide the background and a feeling for the times, and in that sense they all speak in this chapter.

The women I expected to meet, the tough, hardened individuals who had fought in trenches alongside men, simply did not appear. Instead I met women whose passions and sensitivities reflected a real horror of war, who cried through much of our time together and who remembered death, not glory.

War came to the Soviet Union one quiet summer Sunday morning, on 22 June 1941. To the people living in the cities and villages near the western border it seemed to descend out of nowhere despite the drawn-out and bitter negotiations that had lasted until the final hours.

To passengers aboard a train, nearing the key crossroads station at Smolensk, it came in a blast from the sky. It was so sudden that they were still sipping glasses of hot, sweet morning tea as the first explosion rocked the cars. En route to visit her older brother for the summer holidays, a fair-haired girl of fifteen was one of the only people in her carriage to survive the blast. Forty years later, sitting with clenched fists and choked voice, Katyusha Mikhaylova described the devastation: "There was chaos and screaming. I lifted my head and saw bodies everywhere, glass and blood. I looked for the little girl who had been sitting with me. I found her body still trembling, but her head had been blown off. . . . I stood there. I couldn't cry. I couldn't speak. I didn't know what was happening."

The tiny girl stumbled out of the wreckage and followed the handful of survivors who began walking towards the city. As they moved in stunned silence towards the fires that already blackened the skies above Smolensk, they had no way of knowing that this was not an isolated incident, that these were just the first bloody hours of a war that would last for four years and alter their lives. Along the two-thousand-mile border, Soviet armies crumbled and retreated as Operation Barbarossa began.

It should have come as no surprise that Hitler broke the Russo-German pact once he no longer had anything to gain by it, having conquered or gained control over most of continental Europe. He envisioned the huge resources of the Soviet Union fuelling his war machine, and the vast territory providing necessary space for resettling Germanic peoples and expanding his empire. Hitler hated the Slavs almost as deeply as he did the Jews, and his great fear was Communism. He planned for the extermination of thirty million Slavs after his conquest of the USSR, which he anticipated within six months.

Stalin, failing in his attempt to plea bargain with Berlin, could do little to prevent the chaotic collapse of his border armies, who were ill-equipped, unprepared and, worse, suffering the crippling effects of his recent purges. The first Soviet cities fell like dominoes — Minsk, the capital of Byelorussia only six days after the war began, and Kiev, capital of the Ukraine, within two weeks. Refugees crowded the roads and train stations while Stuka dive bombers swooped down on the masses, and Messerschmitt crews machine-gunned them from the skies. German forces advanced in three directions: to the north, towards Leningrad and Moscow; to the south towards the Crimea and the Caucasus; and due east, into Byelorussia, the Ukraine and the Don Basin.

In Heyko, a small town in western Byelorussia, mass evacuation began immediately. Sophia Kuntsevitch, a pretty student teacher of only sixteen, knew there was no possibility of returning to her family's village, so instead she headed with a friend for the Ukraine. "You can't imagine what was happening at the train station. Small children were screaming, people were shouting. I was in shock. Everyone spoke of war, but I didn't fully understand what war meant. During the night our train was bombed. Everyone got off and began walking."

Everywhere military headquarters were chaotic, crammed with people trying to enlist, while thousands more stood lined up outside. In the collapsing republics, formal regulations for volunteers were quickly forgotten as the commanders prepared to evacuate. Young women flocked to join up. Despite official policy denying women combat roles, many officers signed them up, the gravity of the situation dictating disregard of rules and regulations. That a million women volunteered and were accepted into the army was not as surprising as it first seems. In the years immediately preceding the war, the threat of attack from Germany led to the creation of wide-scale, if rudimentary, military training for teenagers of both sexes in the Communist Youth League (Komsomol). In high schools extra-curricular Red Cross courses were jammed with students. So the young women accepted into the army possessed needed skills.

Still, they volunteered out of choice — unlike the men, who were expected to fight — and in many cases did so against their parents' will. The devastating and immediate effects of war with Germany constituted a strong stimulus, and the call to all Komsomol members to enlist was answered in the extremely emotional mood of a reeling and stunned people. As brothers, fathers and boyfriends were killed in the first days of battle, some women swore personal oaths of revenge. Furthermore, a strong historical tradition of Russian women fighters and revolutionaries, known to all schoolgirls, had some influence. But most of all, all the women interviewed for this chapter said that there was the feeling that the war would be over quickly, "and we did not want to miss it." Finally, "We didn't know what we were volunteering for. No one realized what the words 'war' or 'front' really meant."

Katyusha, the young girl who had been on the bombed train, found her way to Smolensk military headquarters. She did not know where else to go or what to do. It was impossible to get back home to Leningrad, and she knew no one in this city. The pale, blonde girl with an almost poetic look was the youngest of three orphaned children, and she was not as delicate and fragile as she appeared, having coped from infancy with life in an orphanage. But she looked even younger than her fifteen years. When she went to enlist and produced her Red Cross certificate, an exhausted and beleaguered recruiting officer almost yelled at her to go away, "'You belong in kindergarten,' he told me." For the next two weeks she helped at a nearby hospital where she could get some food and a place to sleep. Then, with Smolensk under attack and the armies evacuating, Katyusha found a commander of a rifle unit desperately short of medical personnel. "He gave me a soldier's uniform. The shirt reached my knees. I was also given a rifle, made in 1930, which was almost twice as tall as I was."

Sophia volunteered when she reached the Ukraine. "I was already trained in explosives in the Komsomol and had taken courses in first aid. But they wouldn't take me. I was small and very thin. They said I was too young.

"I went outside the headquarters, where there were thousands of teenagers hanging around. I was crying. A Komsomol boy came up to me and asked me what the matter was. When I told him I was too young to join, he took my identification card and changed the date of my birth to make me two years older. I waited for three days and returned. This time I covered my hair with a scarf and hoped they wouldn't recognize me. It was the fourteenth of July, and the situation was critical. The army was preparing to evacuate. Everything was a mess. this time they paid no attention to me. I got assigned to a mortar regiment. On the nineteenth of July I swore the military oath and got my uniform.

"I looked like a real clown in it. The boots were six sizes too big and the trousers were enormous. The soldiers started to laugh when they saw me, calling me scarecrow. Of course, these are petty things," she sighs. "The worst part was that they cut off my blond curls. I had very nice hair. I wouldn't let them cut it all; I left a little bit tucked under my cap. I wanted to look pretty."

Sophia came from a poor Byelorussian family of nine children. Her father had been killed in an accident before the war. Her mother, a Pole, was a religious woman who taught her children to be generous with what little they had. Sophia describes herself before the war as a "real little greenie" in life whose interests were dancing, singing and gymnastics. She finished school two years early and left home to become a student teacher at only sixteen. She says that her athletic abilities and the basic values she learned from her mother gave her the solid foundation to help her in the years that followed. Like Katyusha, she would not learn the fate of her family until the war ended. Both said it was one of the greatest burdens to bear. Family members separated in the convulsive first days simply lost touch with one another. The post and telegraph systems ceased to function properly and in any case were need by the military.

Katyusha's regiment proceeded north from Smolensk to join the defence forces of Moscow, which came under attack in October. There was no time to adjust to military life in retreating armies. She threw away all unnecessary equipment, including her gas masks, but could scarcely carry the pack, gun, shovel and medical bags on the forced marches of thirty miles a day. "My boots didn't fit and I tried to wrap cloth around my feet to keep them on and prevent my feet from bleeding. We slept in the open, in ditches or trenches that we dug and covered with branches. In the bottom we put chips of wood, anything we could find. We slept together, fully clothed, in case the bombing started again."

Katyusha described her first impressions of battle: "All you experienced was confusion, chaos and fear. Explosions, shattering noise coming at you from all directions. You were so disoriented it was hard to tell where the enemy was or where to go."

As Moscow came under attack, the Soviet air force needed pilots and aircraft desperately. Marina Raskova, a pilot famous for her world record non-stop flight to Asia, proposed the formation of a separate women's air regiment, tapping the unconsidered resource of thousands of women who had learned to fly in the sports schools before the war. The Defence Ministry agreed and asked her to form three regiments — fighter, short-range bomber and night bomber units — composed solely of women.

One of the first pilots to volunteer was Nadya Popova, a nineteen-year-old aviation instructer in Donetsk, Ukraine. An outgoing person who then saw life as "amusing and interesting," Nadya had also learned to parachute, without telling her parents, who learned about it when a local newspaper ran a story on her. "I was a very lively, energetic, wild kind of person. I loved to tango, fox-trot, but I was bored. I wanted something different. Most of the girls I knew didn't want to learn to fly. They were frightened, especially of jumping. I don't know why, but it never frightened me." The chestnut-haired, blue-eyed young woman knew how to use her charms. She talked friends into lending her enough money to get to Moscow, where she presented herself to Polina Osipenko, one of Russia's famous women pilots. "I told her how badly I wanted to become an instructor. She helped me enter the aviation school in Hershol, Ukraine. When I got there the male instructor told me I should go home, get married and have children!" By the time the war broke out, Nadya was teaching men to fly.

"The day after the war broke I joined everyone else trying to enlist. I was flatly refused; they weren't taking women pilots. When I got the letter from Moscow a few months later, I immediately volunteered," she says, but adds, "Honestly, I didn't really know what I was volunteering for. I didn't know what the word 'front' meant. I was the type of person who used to be afraid to cut off the head of a baby chicken." But the war had already engulfed the Ukraine. Villages and towns were strafed and bombed, and in the first two weeks her younger brother was killed. "The trauma turned me into an adult in a matter of moments. The change was astounding. I couldn't sleep. I was in shock. My mind had to deal somehow with all the horrible, simultaneous tragedies. It was hard to believe how much grief war could bring so suddenly. When the German planes few over us, we could see the smiling faces of Nazi pilots while they shot into the crowds, gunning down the women and children fleeing to the east, with their bicycles, cribs, bundles on their backs. I couldn't believe what I was seeing. I couldn't believe life could change so utterly in a few weeks."

Nadya left her parents in an "extremely emotional state" wondering if she would see them again. She travelled by train to a military air base near Saratov on the Volga River, at Engels. A thousand women converged there, moved into the long barracks and began adjusting to military life. Training on three classes of planes lasted fourteen hours each day, both in the cockpit and in the classroom. The young women were divided into groups based on their abilities. Nadya, who had flown light sports planes, was directed to train on the U-2 light planes which would be the staple of the night bombing regiment. She trained in a closed cockpit during the day, learning to fly without seeing the ground, a necessity for flying night missions over hostile territory without radio contact.[2]

The three-month training period began with distribution of male uniforms and orders to cut their hair. "I remember one girl with gorgeous long hair crying, saying, 'Long hair is a woman's treasure, her beauty.' It was our first taste of military discipline." In an ironic twist, an air force general, arriving to inspect the women weeks later, lifted one woman's cap off, and the long hair she had refused to cut came tumbling down. There was silence on the parade ground. Then he said, "Now that is what a woman should look like. Why do all of you want to look like boys?" When they got to the front, most grew their hair back.

Military life was a difficult adjustment for some women, hardly more than children, who had never been away from home before. "But your evaluation of things changed," one woman told the author.[3] "What seemed absolutely horrible in the beginning gradually became more or less acceptable. Later, when new women arrived at the front and experienced the hardships for the first time, we would wonder what they were upset about. We were used to not washing for weeks, to the lack of privacy, to the embarrassment of lice inspections. Under fire for the first time new recruits would become hysterical, while we had become calm about it."

Personalities changed too. Some enjoyed the relative freedom, the new responsibilities and "adult" life. "At first I wanted to act like a man," the author was told. "I flattened my breasts under my uniform and learned to swagger when I walked. I learned to swear and how to psych up before a battle. I began smoking, and the most cherished thing became a piece of newspaper and some tobacco. Cigarettes had such a calming effect on the nerves."[4]

Others found it difficult to adjust to the discipline of military life. "Orders had to be obeyed immediately, and there was no relaxation of rules for women. It got very annoying to have to salute a stick of wood ten times, and if you didn't do it correctly, ten times again! It was very, very strict."[5] At Engels the women pilots experienced less military discipline than did women in the army, but there were practices in marching and shooting.

The real test of character came in combat. Nadya, assigned to the 588th Night Bomber Regiment, flew her first combat mission near the southern front in the Ukraine. Women mechanics checked the light plywood planes, remodelled as bombers by strapping bombs to the wings. The bombs were released manually. A pilot and her navigator comprised the crew. One by one, the slow-flying machines took off. "It was a very, very dark night. Not one small star could be seen. The sky was covered in cloud; it seemed that it was an abyss of darkness, pitch black . . . and when I got up in the air, I could see the front line marked by green, red and white tracer lights, where skirmishes continued throughout the night. I followed the lights towards the accumulation of enemy troops. Suddenly, the plane in front of mine got caught in three and later five projector lights, which blind pilots. I watched them fall to earth right in front my eyes and saw the explosion of flames below. I flew towards the enemy lines, thinking I must help my friends. Irrational thoughts . . . I knew they were dead. We dropped the bombs on the dots of light below. They shot at us and I circled round and flew back towards the base.

"When I landed I could see they already knew. I was ordered to fly another mission immediately. It was the best thing to keep me from thinking about it." In stunned grief Nadya continued flying until dawn. That afternoon, the women gathered to mourn for Liuba Okhovsky and Vera Tarasova, the first of thirty-three women who would be killed in action in the 588th Regiment.

"The first time in battle I was very, very frightened," says Sophia, whose unit was retreating eastward in the Ukraine. "The bombs started to fall and I started to dig immediately. I only dug a few shovelsful and flung myself down and lay there with my arms covering my head. I was petrified. When I finally looked up it seemed as though the planes were right on top of me and each bomb would fall directly on my head. I noticed a large crater where a bomb had fallen, and I crawled over to it. I hid myself there. The fear paralysed me, I didn't think I would be able to move. I tried to pull myself together.

"As soon as the planes left, I heard cries of 'Sister!' coming from every direction. There were so many wounded men. I ran up to one. I was feeling sick, I didn't want to touch him. He was completely covered in blood. I didn't know where to begin. Then I just began, automatically, to bandage him, then another. A new wave of bombing started, but by now I was occupied, so it was not as frightening. The only thing that was going on in my head was that there were so many more to help. I kept thinking, just one more, just one more. I saw terrible wounds, jaws torn off and worse. I crawled through the mud, already soaking with blood."

In the first months of war, the combat was of a kind previously unknown, as retreating Soviet armies threw every conceivable kind of weapon into battle, in some cases fighting on horseback against the orderly and technologically superior enemy. As fall turned into an early winter, battles were fought in the most inhospitable terrains, in the worst kinds of weather. In the fierce battles near Moscow, tanks stuck in the mud and sank into the marshlands. Battles left the landscape smouldering until snow came and covered the steel and the bodies. In one of these battles Katyusha was hit by a high-explosive shell. "The airwave lifted me up in the air and took off the shin bone and tibula of my left leg." She regained consciousness in a field hospital only to be told that her leg must be amputated because she had contracted blood poisoning from her wound. The fear of losing her leg prompted Katyusha to consider suicide. "I was not even sixteen. I couldn't imagine living like that. I knew I couldn't survive the war with one leg." She asked the medical assistant to help her get some rope. Instead he told the doctor, who promised her he would not amputate. "He came to me and said, 'You are young, maybe you can fight the blood poisoning.'" She recovered sufficiently to be evacuated to the rear, to a sanatorium in the coastal city of Baku.

Two months later, in December 1941, Sophia was also wounded. "It was during the night. We were trying to dig a trench to put the wounded in. We were in an open field and had no place to hide. I looked around for something to cover them with, a discarded coat, some corn stalks that would serve as a mattress. I wanted it to be ready when the next attack began. The first firing started and I crawled toward the men. I found one wounded and began pulling him back toward the shelter. It was difficult, all that dead weight: a soldier seems much heavier when he is wounded. I put him on my cape and pulled it with both hands and even with my teeth. I was crawling along the ground this way, and suddenly it seemed as though I was surrounded by fire. When I tried to stand up I was hit in the stomach. I didn't really understand what had happened to me, and I stood and began to run. Then everything went black.

"It was a very severe battle, and I don't know how they carried me, through the night, to a field hospital. I didn't regain consciousness for a long time. When I did my unit had moved on and I was kept there to recuperate for three months."

As soon as she was released from hospital, Sophia reported to the command post. She was assigned a position at the field hospital but argued, saying she wanted to go back to the front. "They said, 'Why don't you stay here? You have been through enough already.' Because so many nurses were killed or wounded in combat and not replaced, many combat units had no one to help in the field, where simple first aid could save so many lives. I was adamant about going back." Eventually they agreed, and Sophia was sent to the front-line units in the southwest Ukraine. But only three weeks later she was wounded again, hit by a bombshell fragment in the stomach and leg. "As soon as I felt better I began to ask if I could go back to my battalion, but the wound was very, very serious, and they wouldn't let me go."

Katyusha recovered quickly in the southern climate of Baku, and soon was up walking with only a slight limp. Knowing she could not rejoin her unit, she decided to volunteer at the naval headquarters. Even after severe wounds both she and Sophia volunteered again for combat duty. Asked why, they said it was a question not of bravery but of necessity. Both were completely cut off from their families and unable to get home. Byelorussia was under German occupation, and Leningrad was enduring a nine-hundred-day siege. They say that the suffering endured by those in their native towns inspired them to keep fighting. It might also have been that they had attached themselves to the military when they had nowhere else to go, and it had come to represent home and family to them.

Katyusha was assigned to a naval sanatory ship in the summer of 1942. German forces now occupied approximately one-third of the industrial areas of the Soviet Union, in which seventy million people lived, but the winter had been costly for them. Thousands died of exposure and malnutrition as lack of supplies became critical. Moscow had not fallen, Leningrad persevered against all odds, and Hitler's six-month scenario now seemed in retrospect an ignorant and impossible strategy. His field marshals wrote that better air supply was crucial and warned against the spreading of troops over too many targets. Against their advice, Hitler moved his headquarters to the Ukraine in July and ordered his armies to advance across the Don to the Volga, where they would split up — half thrusting towards the Caucasus and the other half targeted for Stalingrad.

On 23 August the German Sixth Army broke through the river only five miles north of Stalingrad. In a six-hundred-aircraft attack they killed forty thousand civilians in the first days of the battle. Inside the city the 67th Soviet Army dug in. By early September the city had become the battleground. Civilians fought with soldiers in their streets. German forces could not successfully encircle the city because the Volga River, which runs to its east, served as its lifeline. It was up this river that the naval sanatory ships made their passage. Katyusha: "On each trip we evacuated about fifteen hundred wounded. It was a horror story. We had to pull many of them out of their burning tanks. The moment the ship set sail, the German pilots would fly very low and very slowly machine-gun the upper decks, rewounding the soldiers and wounding the nurses."

On the three-day voyage through the Caspian Sea to Baku, severe storms rocked the ship. "There were about 850 wounded soldiers on the ship. Some were so severely wounded they were completely bandaged, leaving only their eyes and mouths visible. Many had to be spoon-fed. The weather and seasickness made them feel worse. Their wounds reopened, and bandages had to be replaced often. On top of everything else, the two other nurses on my deck were so seasick they could hardly work."

Katyusha was helping to load wounded onto the medical ship when the vessel was bombed. "There were flames everywhere," she recalls. "People were screaming, the wounded couldn't even move to save themselves. The ship was sucked under, the water was full of drowning people. Rescue ships approached as I was starting to submerge in the freezing water. I don't remember much, but they tell me I was rescued when I was near the bottom of the sea. When I was in the hospital, a nurse told me I was born lucky."

Katyusha continued on these evacuation missions until the Stalingrad battle ended in February 1943. Amid the burning ruins of the city people had clung to life by a thread. German soldiers had dug themselves into trenches in the frozen ground and died of exposure. On 2 February, against Hitler's orders, the German commander surrendered and the last sixteen thousand German soldiers were taken prisoner. Seventy thousand were dead. "I saw it with my own eyes," says Katyusha. "When the battle ended I saw the prisoners taken out of the shelters where they were hiding. They were wearing anything they could find. They had robbed scarves and coats from dead women, anything to keep them warm. It was a pitiful sight."

The winter was difficult for the pilots of the 588th Night Bomber Regiment. They flew combat missions across the Don

River, in the foothills of the Caucasus and in the Crimea. With up to three hundred kilos of bombs strapped to their wings they took off an average of fifteen times a night, bombing railways, bridges, supply depots and troop positions that were most heavily fortified with anti-aircraft guns. The planes were unheated and the pilots suffered from frostbite and exposure. "The conditions were very severe," says Nadya. "Some of the planes were open, and during the night the temperature would fall below zero. We were always half frozen, and we got colds and the flu. Today my hands are practically crippled by rheumatism as a result.

"We were frightened all the time. You didn't think about it during the flight, but later. We flew each night, and all through those years never slept enough. There was an enormous strain on the nerves. I would see images of burning planes crashing with my girlfriends in them whenever I shut my eyes. I hoped that if my time came it would be an instantaneous death. The thought of being severely disfigured scared me much more. I was young and even pretty, and I wanted all the things every young girl wants. I wanted to live a life I would not be ashamed of if I survived.

"I knew capture would mean rape and torture. That frightened me more than anything. We carried guns and knew if we were shot down over enemy territory to keep one bullet for ourselves. There was fear, of course, but our training really helped. During each mission there was so much to concentrate on, decisions had to be made in split seconds when the plane got caught in searchlights or had flak coming at it. It was later, when I was trying to sleep, that I would relive each mission and start to imagine what might have happened."

As the 588th moved around, its members often shared bases with men. At first there was discrimination. Their first assignment had been delayed because the male commander of the air base thought their training was inadequate, their inexperience in searchlights hazardous, their discipline lax and their presence a distraction to his men. In the first test of formation flying he had his men simulate an enemy attack, which broke up the women's formation and reflected badly on their training. The women's embarrassment was complete when they discovered that it was Russian pilots who had "attacked" them.

The prejudices evaporated, for the most part, after the 588th had proven itself. In fact, by 1943 damages inflicted by the night bombers were so extensive that the women earned the distinction of becoming a Guards Unit, the highest honour awarded a regiment. It was the first female regiment ever to receive that honour.

On-base life included all the wartime deprivations, but also some pleasures. Besides lack of food and other necessary supplies was the constant worry about families left behind. The women formed close friendships and say that the 588th became their family. On mixed bases romance helped distract from the death and destruction. "We always tried to look good," says Nadya. "We wore a little make-up even if it was forbidden. You only have your youth once. Ours came during the war, but still, we were not going to miss it."

Asked about sex at the front, most of the women, now in their sixties or older, were usually reticent. But the author was told that "there were certainly a lot of front-line love affairs." One woman said, "But obviously it was difficult to have sex. To do so you need the time and the place. During the war we seldom had either. There was absolutely no privacy. But still, girls did get pregnant and were sent away."

"I lost my virginity before a big battle," remembers another woman. "My boyfriend asked me if I had ever known a man. I said, 'Of course not.' He told me he had never known a woman. I know this sounds silly, but we didn't want to die without experiencing it."

"Men and women were shy with each other then," I was told. "Today, it seems impossible to think of a million women with the front-line troops, without a great deal of sex going on. But then we weren't like that. The conditions were hardly conducive to sex anyway. We were filthy, exhausted and hungry. We were just trying to survive. But there were a lot of cases when married men fell in love with girls at the front and never returned to their families afterwards. So we weren't entirely popular with everyone when the war ended."

Lack of privacy was one hardship almost everyone talked about. Sophia told one touching story: "It was winter — cold, frost, snow — and we had come to a field where we could rest a bit. All the men immediately walked off to relieve themselves. When we were ordered to move on, I couldn't move. I just sat there and began to cry. The men asked me what the matter was, and I explained that I couldn't go to relieve myself because there was no privacy, no bushes in sight, and I was desperate. So the men made a circle with their backs to me so I could go. When I finished my hands were so frozen that I couldn't do my pant buttons back up. I couldn't stop crying. One of the older commanders who had always treated me like a daughter came and did up my buttons for me."

The soldiers seemed sensitive to women's needs and where possible tried to make life a little easier for them. "They would help us carry things, when they could, and in return sometimes would ask us to help them wash their clothes or sew on buttons. We didn't mind doing it. They did things for us. Life is not easy for anyone at the front, but particularly for women it was hard. Not having the same physical strength, and things like menstruation — those four days a month became an enormous inconvenience during the war. But we had a different kind of endurance, maybe because we knew we didn't have to be there. So we became that much more determined.

"There were strong friendships between us all. All these feelings were tested in extreme conditions. Because at any moment you could be killed. The war lasted for 1,486 days. That is how long there was a possibility you could be killed. So the protective feelings were immense. I think women's presence at the front had a dramatic effect on the soldiers. They tried to look presentable, they managed to shave themselves and not to swear too heavily when we were around. It brought out their gentle side, when normally they would have been brutal. Sometimes they would pick field flowers for us.

"I gained a new respect for men — soldiers — born not out of idolatry from afar, but out of sharing this with them, exposed to their weaknesses, seeing how they coped and showed more human sides. They cried, they were frightened, they were upset about killing. It was important for me to experience that, alongside them, as their companion. Some were artists, musicians, the very special people who should not have been sent to the front but who always seemed to be the first to volunteer. It seemed that the best and most talented were the first to die. That left a great impression on me."

During the war women were proving that in some areas they were not only capable but sometimes more adept than men. One such area was patience; the best snipers were women, because they could sit absolutely still for longer periods of time, and stay disguised. A women's sniper school near Moscow trained 150 women every two months during the course of the war.[6]

The navy was the most traditional and chauvinistic service and the most resistant to allowing women into combat. When the battle for Stalingrad ended Katyusha left service on the medical ships and answered a radio announcement for volunteers to a newly formed marine unit. "When I applied the commander just laughed and said, 'If Peter the First [the founder of the Russian navy] heard that you wanted to enlist in the navy, he would turn over in his grave.' I got quite angry and said, 'Well, whatever Peter the First thought is history. I'm a Soviet girl, and we are equal in this country.'

"I knew I would never get anywhere with him, so I decided to write to Naval High Command in Moscow. I told them I had enlisted when I was only fifteen, that I had taken part in battles near Moscow and been wounded, that I had served almost a year as a nurse on a sanatory ship. Now my native city Leningrad was under blockade, people were dying of hunger in the streets and I wanted to defend my country.

"The commander received notice from Moscow that I should be allowed to enlist. But when I arrived on the ship the captain had told his sailors that here was a girl who was good for nothing and crazy enough to desperately want to join the marines, and told them to give me an appropriate welcome. One of the sailors walked up to me and handed me a baby's pacifier. He said, 'This is for you. When we go into battle we won't have time to baby you, so take this.' I said, 'We'll see who is going to take care of whom.'"

In her first operation Katyusha joined the armies in a forced fifty-kilometre march, "in field service marching order, in the burning Caucasus heat, carried out partly in respirators." Men collapsed from sunstroke. Katyusha refused to give in despite the pain in her old wound, which swelled her left kneecap.

They received intensive training on ship and on land for amphibious landings, sea and land battles. She remained the only woman in the battalion but underwent the same training as the men and slept among them. She described life aboard ship as "difficult. Most of the problems were of a simple character compared to combat, but the men tried to give me what privacy they could. They were probably as embarrassed as I was."

Katyusha would serve a dual role as nurse and combatant, laden with grenades, anti-tank grenades, a machine-gun and a medical bag. Before each military operation, she says, "There was always fear. I was afraid of every German. Only brainless people are not afraid. But somehow I knew I had to do this. We began our first amphibious landing with artillery fire coming directly at us, into our faces. . . . When we got on land we fought in the streets, in the open, with machine-guns, tanks." Her first marine operation was described by a Soviet journalist: "It was the capture of Temryuk. They made a landing in the flood plain while mortar shell explosions caused pillars of water to rise in quiet backwaters, bullets made water boil and the reeds, cut down as if by an invisible scythe, fell on top of the marines' heads. She was there, in the midst of all this mess, stood in salty water up to her chest, fired and dragged the wounded into boats. It was only a tiny town, but it was paid for with a very high price: more than half the battalion fell there, in the flood plain and on the coast."[7]

"After this first battle the soldiers came to me, one by one, and apologized. They said, 'Katyusha, we are on our knees before you. Please pardon us for being cruel when you joined. We now know that not only are you brave, but you saved our lives, rescued us and carried us from the water, tended our wounds and fought with us. Please forgive us.'"

When asked about her reactions to killing, especially as a medical worker trained to save lives, Katyusha became a little angry and defensive and asked what anyone in her place would have done. "In the front, one cannot be without arms, one has to defend oneself. As a matter of fact, I did not want to carry a gun in the beginning, but I learned it was a necessity. Sometimes I would leave my gun, but I understood later that I could not defend the wounded from the enemy without it. I didn't enjoy killing. It was done out of necessity, to save lives."

Nadya: "I don't think you can separate men from women in this situation. War does not spare anyone, it doesn't distinguish between the sexes, or the young from the old. It was a who-will-win situation. They were destroying us and we were destroying them. There was no choice involved. That is the logic of war: it is life or death, victory or be vanquished. I killed many men, but I stayed alive. I was bombing the enemy — the target — just some lights on the earth below. War requires the ability to kill, among other skills. But I don't think you should equate killing with cruelty. I think the risks we took and the sacrifices we made for each other made us kinder rather than cruel."

Another woman told the author: "I heard men say that the first killing is very difficult emotionally, but after that it becomes automatic. I never felt that way. Each day, with every death, I went through hell. It never became automatic or instinctual for me."

Some spoke of the brutalization, the numbing effect of the war on their personalities. One young army woman volunteered to participate in a firing squad to execute two deserters. Afterwards the men in her unit would not talk to her. She was told that the war had made her cruel. After the war, under psychiatric treatment, she was told to get married and have "lots of children" to restore her soul. One thing seems certain: men were repulsed by women who behaved in a tough or cruel way, even when they did the same themselves. As for the enemy, Soviet women soldiers were considered curious, even disgusting. Early in the war, the Germans captured more than a hundred thousand Soviet women soldiers, whom they called *flintenweib,* a derogatory term loosely translated as "musket-women." German women who served in the women's auxiliary were warned against becoming like the *flintenweib.* It is apparent that no such warnings were issued to the SS women, for whom, working in the concentration camps to which Russian women POWs were sent, cruelty had become an obsession.

More devastating than killing was the effect of the dying and wounded, says Sophia, who by late 1942 was serving with the 261 Division in the Caucasus Mountains. "Seeing the suffering that I could not stop — the dying, those who had lost arms, legs, faces . . . it made a deep impression. At the same time you couldn't show it. I had to appear cheerful, to encourage the wounded, to lie to the ones who were dying and tell them they were all right.

"I felt a great deal emotionally for each wounded person, but when the young girls were killed or wounded it was almost more than I could stand. I remember one young girl. . . . I spotted her sitting with her back against the tree during an attack. I thought she must be crazy, she'd be hit instantly in that position. Every time I moved from my trench, even the slightest distance, I was shot at. I thought she must be in shock. I crawled over to her. She seemed all right. I pulled open her shirt to see if there were any wounds, and there were none. Then I took her face in my hands and turned it towards me. She was very beautiful. I saw a tiny wound, without even a trickle of blood, in her temple. She was dead. I put her head against my chest and began to cry."

Sophia described the battles in the Caucasus as the worst she had experienced. Most of the Russian soldiers there were extremely young and fighting with no experience. In the bitter infantry battles in the mountains, she claims that the Germans were given alcohol before they attacked, then advanced, shoulder to shoulder, in waves towards the Russian positions. "To defeat them we had to leave our trenches, stand up and meet them. It was extremely difficult because the trenches made people feel safe. Seeing the German suicide attacks, watching them come forward with their mad, drunken eyes, it took tremendous courage to go forward. Our oldest men went first, and they were the first to be killed.

"Our battalion consisted of about six hundred people. At the end of one battle there were no more than seventy of us left. We had fought all day long. I kept crawling to the command hill with ammunition and returning with the wounded. On one trip I noticed that the last remaining officer had been hit in the head. I bandaged up his head and dragged him to the hiding place. I tried to give him a little water to drink, but he couldn't. I tried to spoon-feed him the water, then I tried to give him water from my mouth. He died a few minutes later. I was afraid to tell the men that our last officer was dead. I didn't know what to do, I was so scared. We expected reinforcements at nightfall. The Germans continued attacking. So I said, 'Listen boys. Our commander is wounded, and he told me that I should take command. Follow me.' I led them into an attack. I ordered the men to split up into small groups and attack from different directions to give the impression there were more of us. I had been in combat since the war began, and I had seen how the military commanders worked. I had had my education in strategy, so it wasn't that extraordinary for me to do it. My only thought was that if we retreated all the hundreds of wounded would be slaughtered. I knew we had to hang on till nightfall. I shot blindly ahead and never saw a face into which I fired. I was doing it for the boys beside me."

When the reinforcements arrived only forty men, and Sophia, were left fighting. She became somewhat famous after this incident. "People heard about it and expected to see a group of real guerrillas led by a woman who was like an Amazon. When they saw that we were simply teenagers, led by a young girl of seventeen, they couldn't believe it." Sophia was awarded the Order of the Red Banner, one of the highest military honours.

"After three days' rest we returned to the fighting. These were terrible times. Many days we did not get our supplies and didn't eat all day. The worst thing was thirst. Water became the most precious thing. There were times when I thought I wouldn't survive. I'd be carrying a heavy man and a bombshell would explode near us. We would fall apart, and I would lose consciousness. When I came to I would feel so miserable that I would pray to die. I couldn't move an inch, not even my little finger. I would think, 'Whatever happens, it makes no difference, I will stay here until I die.' I don't know where we get our strength in these moments. I don't know how we can take these things. But all of a sudden I would get a force that would come from nowhere, and have the strength to do what I was supposed to do."

She was wounded for the third time shortly after, receiving a contusion caused by an exploding shell. Her arms were so stiff and enlarged, and her shirt so thick with blood, that the nurses had to cut the shirt off her. She fell into a coma, and for many days the doctors did not believe she would live. They transported her to a hospital in Armenia where she finally regained consciousness. In the meantime the men who had survived the battle tried to find out what had happened to her. They spoke to a journalist, describing her and what she had done. Someone in the hospital saw the resulting article, and a few weeks later her friends and a delegation of the Communist Party of Armenia came to the hospital.

"They told me that they admired my courage and that I had saved the lives of many Armenian soldiers. They offered me money to leave the army and enter the medical institute in Moscow to become a doctor. I was very grateful, very touched. But I had to refuse. My native country Byelorussia was still occupied, and I couldn't dream of leaving the army until it was liberated.

"For a long while after this the contusion caused severe depression, which made my recovery very slow and painful. I didn't yet know the extent of human capacities. We can go through so much and still recover. I now believe that ninety percent of different nervous disorders could be overcome by the patient himself if he set his mind on it. That was my greatest difficulty — getting through these mental attacks."

Katyusha received the Order of the Patriotic War for her participation in the amphibious landing and attack in Kerch' El'tigen, which began on 31 October 1943 and lasted until 11 December. The beachhead on the Kerch' Peninsula had to be secured so that troops could launch operations into the Crimea. It was one of the largest amphibious landings made by Soviet marines during the war, and occurred during a storm in the middle of the night. "It was a mined harbour, and there was a wire about seven feet beneath the water, which was mined. Sailors got caught in the wires. I tried to unhook the men and carry them to land. Each sailor weighed about one hundred kilograms, while I weighed only forty-three. But under fire an inner strength seems to appear, and you have enough strength to carry anything — even a man with all his weapons." On the beach hand-to-hand fighting ensued. Germans surrounded the five-kilometre beachhead.

The marines fought without food and supplies for days. The 588th Night Bomber Regiment was ordered to provide air support and drop supplies, and the pilots made the 1½-hour flights with cannisters of food strapped under their wings. Nadya remembers the treacherous missions, made more dangerous by the difficult weather conditions and air current that flung the light planes around. When the planes approached the beach, Soviet troops would open diversionary fire. Nadya: "We flew in low, cut off our engines at about fifty metres and released the supplies. We could hear the voices of the marines below shouting thanks. We didn't lose any aircraft during these missions, which was amazing. I was shot at each night, and landed with holes in my flying kit."

On the beach below, Katyusha's biggest concern was finding water for the wounded. It is said that the Germans learned of her presence there and called out her name, allowing her to cross in front of them to fill buckets of water from a well.

The 588th was assigned to the 2nd Byelorussian front in the spring of 1944. There, in the pastoral fields, farmlands and forests, another face of war revealed itself. The charred remains of five thousand villages, systematically demolished after all their inhabitants had been executed, stood as testimony to the genocidal policies of Nazi occupation. One out of four Byelorussians had been exterminated, with no distinction between men and children, the aged or women. Death camps were less sophisticated than in the west: here people were stacked between wooden stakes and burned to death. Savage partisan warfare raged from forest camps, sustained at great cost by underground agents in the cities. Even some German soldiers, horrified by the realities of Nazi rule, aided the underground, supplying them with guns, informing them when raids were to take place. In the bombed-out shell of the capital city, Minsk, German reprisals were horrific. Thousands of bodies of civilians hung from lamp-posts to decompose. Others were buried alive, their screams and moans audible. Twenty-seven thousand women engaged in active partisan warfare, and many more worked in the underground. Three women resisters in Minsk executed a German general in his bed, and in reprisal a thousand civilians were shot. As the Soviet troops pushed across the Byelorussian front, German atrocities increased. Nothing remained of Sophia's village as Nadya approached the republic by air.

"As I flew closer, over Minsk, I could see a few buildings, the opera house, the government building, a church and an officers' headquarters. The other buildings were completely destroyed. I became very emotional and afraid. I didn't think the people could ever restore this city, which lay there like a gaping wound, smouldering.

"I landed in a forest clearing a little way from Minsk. It was not an airport, just improvised, but there was a rest place for us before taking off on our night missions. As we lay there on the grass trying to sleep, we were approached by a very feeble old woman. Her clothes were in rags. She carried a small basket of berries which she had picked in the forest to feed herself. It was evident that she was starving. You could see traces of it on her face. She offered us her berries. It was all she possessed in the world. She knew we were fighting and she wanted us to have them. She said, 'Please take revenge for me on the Germans. They killed my husband, burned my village to the ground. I have no home, no family, nothing is left. I have been living with the partisans in the forest. I don't know what has happened to my three sons, who left to join the army.' Some children came up behind her and stared at us. They were also living with the partisans because they were orphans. For the next few weeks, every night when I was flying, I saw her face and the faces of those children."

On 3 July 1944 the last German troops were defeated in Byelorussia. Almost four years to the day after it had started, the war in the Soviet Union was over. Now Russian armies moved west, fighting their way to Berlin. Katyusha, Nadya and Sophia continued serving. Sophia recalls that operations with an advancing army were much easier on her and all the medical personnel because the wounded could be left on the battlefield and within hours evacuated to the rear. "But sometimes the trucks were in such a hurry to bring in supplies, they didn't want to stop and pick up the wounded,'' she says. ''I used to stand by the side of the road and step out in front of the trucks with my machine-gun, ordering them to stop. I'd threaten to blow out their tires if they didn't pick up the wounded.''

Katyusha's most memorable battle occurred in Yugoslavia. She talked about it twice, passionately. In December 1944 the marines of the Aziv Flotilla stormed the fortress of Iluk, which stood on a high-hilled island in the Danube near the city of Vukovar. It had to be taken by land, but the marines were ordered to make diversionary landings on an island nearby. There were about a hundred men and Katyusha. The island lay low in the river, and its banks were often flooded. When the landing force arrived that night most of the island lay under water.

''During the landing, first the water covered our feet and then it was over our heads. In some places I dove into it head first, swimming. Germans in the fortress opened fired — machine-guns, tracer bullets, red, blue and green. They were shooting at our foreheads, the sailors were falling into the water and dying. I had to rescue them, they were losing blood and drowning. I didn't know how to help them. I raised my head and saw a half-drowned little piece of land and began to pull the wounded men towards it. I dragged each one up and tied them with my bandages or their belts to the tree branches which were above water.

''The water was on fire, boiling from the machine-gun bullets. Germans surrounded us, and the men told me to blow them up with a grenade, and myself also, because there was no way we could survive. As dawn came the enemy could make out our forms. One aimed for my heart but got my shoulder. The exploding bullet ruptured an artery, and the blood poured out like a fountain. I passed out and fell into the water. When I regained consciousness the Germans were very close — only ten feet away. I took an anti-tank grenade from my belt and threw it at them. Another man, our cook, took several grenades and jumped towards them, dying instantly.

''There was a short silence, but then they began to surround us again, coming from the bushes behind us. I had no grenades left. I began to shoot from the machine-gun. They were shooting at me; they didn't take into consideration the fact that I was a medical person or that we were all wounded. The fighting continued until the afternoon, when our naval forces managed to conquer the fortress. There were thirteen of us left.

"Eighteen hours later an armoured boat came up to our island to see if there were any survivors. We had been waiting for them, bloodless in the icy water."

Katyusha was taken to a hospital in Novisad, where twenty-two fragments were extracted from her shoulder. She had also contracted pneumonia. "It is still very difficult to talk about that battle," she says. "As I speak, I am living through it again, and it is very painful. Those men who died were all my friends."

Sophia arrived in Berlin in the first days of May. With two friends she walked to the centre of the city and carved her name on a pillar. "I wrote that I, Sophia Kuntsevitch, Russian daughter of a welder, came here and defeated fascism." But in the mop-up operations that followed, only five days before the official victory, she was hit by a bullet in her liver.

Nadya was among twenty-three women from her regiment who became Heroes of the Soviet Union. This highest honour was conferred on her just before her final mission to Berlin, where she rendezvoused with a fighter pilot she had met early in the war. The two pilots were married there. "It was there I was told by German soldiers that they had called us night witches. We women had a reputation. They told me that they thought the night witches were horrible looking, that we were given injections or pills to see well at night."

Katyusha was in Austria on 9 May 1945. "For four years we had slept out in the open, but this night we were billeted in one of the military premises, in a five-storey building. I was sleeping in a room on the top floor. At about three in the morning I woke to the sound of shooting. I grabbed my gun and a sanatory bag and ran downstairs and out into the courtyard. There I was amazed to see my friends, my comrades, kissing Austrian women. I said, 'Are you crazy? There is shooting somewhere!' They screamed and shouted at me that it was all over, that we had won the war. I began to cry. I threw down my gun and said, 'I will never pick up a gun again.'"

It was all over and the soldiers now faced the long, difficult journey home to the places of their youth, preserved unharmed in their memories. But there were no home-coming parades, no time even to mourn the dead. They returned to destruction, burned-out remnants of what life had once been, anxious to learn the fate of family and friends.

Sophia spent all of 1946 in hospital. Again, she was not expected to live, but with the same fortitude and strength of mind that had seen her through her other injuries, she slowly became better. "It was an awful year," she says, crying. "I learned that most of my family were dead. I didn't have nightmares, but I thought about the war constantly. I could not get rid of the images of the battlefield."

Nadya returned to the Ukraine to search for her family. "I didn't know if they were dead or alive. Through the war we had heard terrifying stories of what Nazis were doing in the occupied zones. But somehow my parents survived the war.

"I felt sick when I saw our old house, which they had used as an administrative building. They had destroyed it, cut down the trees, and what used to be gardens was just mud. Down the road the school I had gone to was bombed. Nothing was the same.

"We had dreamt of victory for four years. Here it was, but not our dreams of returning to a good night's sleep on clean sheets, to life unaltered. We came home to face all the destruction and severe food shortages. We worked eighteen hours a day to reconstruct. Maybe that is why we didn't have much post-combat stress — we didn't have time to reflect on our personal experiences in the war, we were too occupied by the present. Most of us got married immediately and began to have children. Only later would I sit and think of our regiment, of the closeness we shared, and become sentimental. The amazing thing is that the close friendships remained. Every year, at least once a year, we have reunions. Our regiment, in total, flew 24,000 missions. I flew 852 myself. It was a lot to go through together — and to survive."

Katyusha Mikhaylova was recommended for the title of Hero of the Soviet Union after her participation in the battle of Iluk, in which she killed fifty-six Germans, but the generals in Moscow refused her. Her commander who had sent in the recommendation later said it was turned down because he had failed to convince them that what he wrote was true. Even in the war in which a million women participated, thousands killed or disabled, the deeds Katyusha's commander attributed to her seemed implausible to the powers in Moscow.

After four years of war the nineteen-year-old girl said goodbye to the men she refers to as "my brothers, my closest friends" and returned to Leningrad.

"It was then that I discovered that all our family had died during the blockade. My brother and sister, both in the army, had been killed in action. I was alone.

"I believe I changed during the war. I became tougher, stronger. But not because I killed men and fought. It is because of what I saw, the death and destruction. When I learned my brother and sister were dead, I had to be strong or I could not have borne the pain.

"The problems of coping are hard to describe. Leningrad had suffered a siege of nine hundred days. One million people had starved to death inside the city. After the war there was no food, no clothes, nothing. The people there had suffered as we had at the front. Women and children knew all about suffering. The final battle that we faced together was hunger."

After the war Nadya was asked by a British journalist if she missed the flying missions or was ever bored. "I didn't even know how to respond. How could I explain to her what we had been through, and that peace was the only thing we cared about? Not one girl in our regiment chose to remain in the forces. We just wanted to return to a normal life."

Sophia, too ill to attend medical school, studied library sciences instead and married a partisan who had fought in Byelorussia. When she was finally released from the hospital the doctors warned her that she must never attempt to have children. It came as no surprise to me, however, when she grinned and pulled out the snapshots of her daughter and grandson.

ENDINGS AND BEGINNINGS
Palestine, 1945-48

We trained with the boys, slept in tents with them and fought with them. The unusual thing was that it was not unusual. We all did it.

Yaffa,
Palmach soldier

I am sure many of [the commanders] would rather have made love to me than send me out to fight. But they had no choice. It was total war.

Shifra,
Haganah fighter

As one historian wrote: "The Second World War was the centrepiece of the last act of the story; the generation after the war was kept busy with the epilogue."[1] That epilogue saw the World War end and the Cold War begin, the defeat of colonialism and wars of independence in various parts of the so-called Third World. It also saw the creation of the state of Israel and the displacement of Palestinian Arabs, in an area of such strategic importance and ongoing tension that it has remained at centre stage ever since.

The horrifying revelation of the Holocaust at the end of the Second World War was the most visible catalyst for the events that followed in Palestine. Although Zionism was an ideology that had been embraced by pioneering Jews in Palestine since the late 1800s, it was the tragedy of the Holocaust, in which six million Jews lost their lives, that propelled the events that followed.

The idea of Jewish women fighting in Palestine was not new in the 1948 War of Independence. An underground army, the

Haganah, had trained women in armed defence since the 1920s. Women, however, were trained for defence of their isolated settlements, not for aggressive warfare. Girls learned basic weaponry in their schools, and youth organizations taught both sexes the rudiments of fighting. By the end of the Second World War Jewish women in Palestine were members of the underground defence organization, and not unfamiliar with weaponry.

Each of the four women I interviewed for this chapter chose to fight for Zionism, yet they chose significantly different ways in which to participate. In that respect their stories give some indication of the differing beliefs and actions of various Jewish groups in Palestine in the early 1940s. What they shared was a vision of an independent state of Israel and the willingness to resort to violence to achieve it.

Whenever one raises the subject of women fighters, people usually refer to Israel. Over the years photographs of young Israeli women in fatigues carrying Uzzi machine-guns have reminded us of the legacy of women fighters that began in the first days of Jewish immigration to Palestine. Over twelve thousand women are said to have fought during the War of Independence, yet it was the last war in which Israeli women were combatant. Today women are still drafted for service in the army and trained in weaponry, but they are not allowed into battle. There have been exceptions to the rule, and there are small groups of men and women who are sent to guard isolated kibbutzim today. But officially women are not permitted to be in combat. When the underground army became official with the Declaration of Independence, a women's corps was formed. Some say the regulations banning women from combat were adopted as a result of political pressure from Orthodox groups, others that they were imposed because women soldiers had higher casualties than men in 1948. Whatever the reasons, there is little doubt that Israel, like most countries, can bear the idea of women in combat only in times of national emergency. Otherwise, it has remained traditional in its defence regulations.

Now in their late fifties, the four vital women I met, whose experiences thirty-five years ago seem to them almost unbelievable, still feel that their involvement in creating a nation remains the most significant and meaningful event of their lives. Only one of the them, however, would permit her full name to be used. I was told by the others that they felt they had done no more than what all young women were doing at the time and that they did not want to be singled out.

Hanna Armoni resides in Tel Aviv today, where she heads a museum of a former guerrilla organization called the Lehi, or Stern Gang, that operated against the British in Palestine. She agreed to meet me and tell me about her experiences from 1943 to 1948 because "I have a great sense of history. I feel we must talk about these things." As a member of a guerrilla group that was extremely controversial and unpopular even among most of the Jews in Palestine, Hanna says that she needed to be strong in the face of solitude. Today she exudes a strong-willed personality and a great deal of energy.

I travelled to New Jersey to meet Yaffa and Shoula, both former members of the Palmach, the fighting arm of the underground. These two women had been best friends since childhood in Haifa and after thirty-five years, by coincidence, both were living in the United States. I interviewed them together. They had not talked about those days for a long time, they said, and since they hadn't seen each other in so long there was almost the feeling of a reunion.

Shifra resides in Toronto, where she has lived for a number of years, but she grew up in Jerusalem and participated in a defence unit there during the siege of 1948, in which she fought in the trenches with the men. She is a highly energetic woman, always busy with community work and family. Although she agreed to talk to me she soon confessed that she tries never to think of the war "because I have too many unhappy memories." Her attitude towards her actions of thirty-five years ago: "I did what I had to do. If I was not there, someone else would have done it for me."

"If you had seen me then you would never have believed I was a 'guerrilla.' I was so thin, so delicate, I didn't look as though I could hurt a fly," smiles Hanna Armoni, a former member of the Lehi or "Freedom Fighters of Israel," which operated against the British in Palestine until 1948.

Hanna was born in Germany and was only six years old when Hitler came to power in 1933. "It was an immediate change," she says. "I was only a child, but I can remember the dark shadow falling." Anti-Semitism, now publicly encouraged and organized, was quick to affect even a schoolchild's life. "The children started singing anti-Jewish songs," she remembers, adding that she felt frightened and ostracized. The following year her parents took their three children and fled to Belgium. When Hanna was nine they arrived in Tel Aviv, Palestine, to begin a new life.

By the age of fourteen Hanna had become a passionate Zionist and "an extremely sensitive girl." She began keeping a diary in which she wrote "not about personal things, but about the world. I wondered what I could do to prevent wars in the world. What could I do to change the world and make it more peaceful? I thought that a personal example would be the best way." It was 1940, and the European war had become a "dark backdrop" to her thoughts and feelings. But Palestine was the centre of her world and by 1943, at seventeen years old, she had joined the Lehi, which was known to the outside world as the Stern Gang after its founder, Avraham Stern.

The adolescent Hanna had wanted "peace in the world" but by seventeen she was ready to fight and use violence to create an independent Israel. She joined an organization that had effectively declared war on the British in Palestine, that was widely unpopular among the Jews and that utilized tactics of sabotage and assassination. Hanna says, "The used of the word 'terrorism' had a different meaning than the one it has today. The Lehi's actions were restricted; we did not hurt anybody who did not directly represent British rule."

The Lehi advocated that the time had come to create an independent Jewish state. Like other Lehi members, Hanna believed that the British, who had ruled in Palestine since 1917, had no interest in giving it up.

During the Second World War most Palestinian Jews had joined the British to fight Hitler. The Lehi maintained that Britain was the enemy and had stabbed them in the back. In 1939, in need of Arab support against Germany, Britain had issued a

White Paper that put a ceiling on Jewish immigration to Palestine. Zionists believed that it revoked promises to help the Jews create a sovereign state. There was a further moral issue: curtailing Jewish immigration to the country would leave millions of European Jews victims of Nazi Germany.

Hanna joined the Lehi after she accidentally met a couple of members. She believed that "we needed to get the British to listen to us. They had given in to Arab pressure regarding Palestine. We believed that we had to make a lot of noise too so they would listen to us. Of course, most of the population thought the Lehi was wrong until the end of the war."

Her parents, "though not in opposition to my beliefs," did not want her involved in the underground. "I was brought up like a princess. I was sheltered," says the blue-eyed, fair-haired woman who was then a slender teenager. "They didn't throw me out of the house, but they weren't happy about it. They were afraid for me."

By the time she joined the Lehi, Avraham Stern and many other of the founding members had been killed by the British authorities. The organization was internally wounded and externally without support. Her first activities were to put up posters around the city and distribute pamphlets that articulated Lehi views, "to try to gain some public sympathy." Most teenagers in the country joined youth groups of the Haganah, a clandestine defence organization established during violent Arab-Jewish clashes in the 1920s. But with the outbreak of the Second World War the Haganah directed its members into the British Army. "All my school friends joined the Haganah," she says. "In order to join the Lehi I had to cut myself off from them. It was a very lonely decision.

"There were quite a lot of women in the Lehi, which then numbered about two hundred people. I struggled for . . . feminism. I believed that a woman could do anything a man can do. I believed that if men could fight in battles, then so could I. The reaction to that was, 'Yes, all right. But you have to wait.' They said it was a matter of experience rather than sex. They said when the time came I would be chosen to help. Some men treated women as equals, some didn't. When I first joined I believed women could even do things requiring a great deal of physical strength. I said, 'I'm strong enough.' I was so determined."

She took the name Sarah for her underground name. "It was the name of a Jewish heroine who lived during the Ottoman Empire. The Turks caught her and tortured her, but she wouldn't talk. She asked if she could go to the bathroom, where she had a small revolver, and she shot herself in the mouth. She was a great symbol for me."

Hanna's first direct involvement with violence affected her deeply. "I thought, 'What right do I have to kill someone? To play God?' It was a moment of doubt. But I didn't think there was any other way. We only killed the British, and a few Jewish traitors. If we saw one with his family, we were very careful not to harm them. Many times members of our organization were hurt or killed because they had avoided harming innnocent people."

One of her first sabotage missions was to blow up a railway bridge near Akko to disrupt British transport of military supplies. "I had been taught to handle explosives and guns. We learned these things in a house we rented in a village.

"We planned the action carefully. One group was to blow up the bridge. My group of three people was to set off explosives farther down the tracks. As in all operations, I was scared. As we left that night I was trying not to shake. My motivation was so strong that it helped. My job was to tie the explosives on the rails. Then we hid in the bushes and waited to hear the explosion on the bridge before lighting our fuses. But just then we heard an unexpected train coming towards us, so we rushed down and untied the explosives. We didn't want to accidentally blow up a passenger train.

"When it had passed, I tied them again. As I was doing so, I noticed that one of the fuses had come out and the bomb could have exploded at any moment. Luckily nothing happened. We lived by luck, too. Finally we heard the explosion on the bridge and lit our fuses and rushed off. Later, when we learned that everything had come off just as we had planned it and that we hadn't hurt anybody, we were extremely happy."

Hanna's main job during 1944 and 1945 was to help recruit

people into the Lehi. "I had to explain to people who wanted to join exactly what we were trying to do. We didn't take any people who didn't completely understand the cause or who weren't sure they could do it." She describes those who did join: "After one minute, strangers became friends. Our common ideal was so strong. Now it seems so rhetorical, but at that time it was real and everybody was a real friend. We understood each other. Everybody was ready to die for each other and for the cause.

"In the underground, though, I learned people have their weaknesses, too," she adds. "You think that in the underground where people are willing to sacrifice their lives, there wouldn't be much selfishness. But somehow there were people who were selfish too. We spent many times holed up somewhere in uncomfortable situations because we had to protect our secrecy. Once we spent several days and nights in a very small, three-room, old wooden house. There were fourteen of us and we couldn't leave it during the day. At night we could go out, but only in pairs. One guy took the supply of cigarettes, which were supposed to be for all of us. We had to ask him each time we wanted to smoke. You can learn the character of people quickly in those situations. He may have been a good fighter, but that didn't make him a perfect human being."

Many sides of human character were being revealed simultaneously as the European war finally came to an end and the devastating truth of the Holocaust was exposed. Of the eleven million victims, six million were Jews, and most Jewish families in Palestine learned that they had lost relatives. Hanna's parents lost their entire families in Poland and Germany, with the exception of one brother.

"To be honest, though we believed that something awful was happening to Jews in Europe, I don't think I really believed it was anything like the Holocaust. We in Palestine couldn't believe it. It was beyond our imagination that people could do that to other people. Jewish people were used to hearing about pogroms and killings — many Jewish pioneers in Palestine had come to escape anti-Jewish pogroms in Poland and Russia — but never had there been anything in such an organized manner. Maybe if people had really believed it it would have caused such a wave of reaction that things would have been different. But we didn't believe it. It was written about in the newspapers, but always with a question mark. And they were being killed, day by day, by the millions, and nobody believed it. Nobody helped them. When the war ended, people woke up from their illusions."

The Haganah and another underground movement, the Etzel, or Irgun, led by Menachim Begin, now came to an agreement with the Lehi to fight the British. Even after the Holocaust Britain refused to let refugees into Palestine. The few survivors who managed, against all odds, to get there in broken, decrepit ships were driven back out to sea by enforcers of the British blockade. Some boats sank with their pitiful passengers aboard.

All Arab nations rejected the tenets of Zionism and the formation of a Jewish state in Palestine. They applied a great deal of pressure on Britain to stop all Jewish immigration, which now was set at ten thousand people a year. While European Jews waited for over two hundred days in the displaced persons' camps of Europe for an Allied decision as to their fate, the Haganah worked to get them into Palestine by clandestine means.

At the end of the Second World War, two school girlfriends from Haifa joined the Haganah's elite fighting army, the Palmach. They were in many ways a product of their generation. As Sabras, Jews born in Palestine, both were deeply committed Zionists, but they were more pacifist in their views than Hanna. They believed in expanding Jewish settlements in Palestine and still hoped that the British would eventually negotiate on the creation of a sovereign state. Like all young women who joined the Haganah youth groups, they had been taught the rudiments of weaponry and self-defence. But when the World War ended they chose to join the Palmach as full-time volunteers. Shoula and Yaffa had been friends from childhood. They shared the same upbringing and the same idealism.

Shoula: "We very much wanted to finish school, but we felt the need to fight — not only for the creation of Israel, but for the refugees who were not allowed in by the British." She explained that the Palmach's philosophy was not to attack the British, but to protect settlements and help refugees. "We did not fight for fighting's sake. Only when we were attacked."

Shoula's father had been a pioneer to Palestine early in the century, and she says that her parents were "proud of me for making that choice." Yaffa's parents were also proud. "It was not difficult for them to accept girls' doing this," she says. "You must understand that all young people believed that it was not a matter of choice, but of duty. We were educated that way — in school, after school, in the kibbutz. We were taught to do our duty because it was our duty."

The Palmach, which means "striking companies," grew out of all-Jewish units who had fought with the British during the Second World War. When the war ended they went underground (since it was illegal for them to form an army) and stationed themselves, undercover, on kibbutzim. During the war several Palmach women had served behind the lines in Europe. Hanna Senesh and Haviva Reich were parachuted into Yugoslavia to help organize escape routes for the underground. Both were subsequently caught and executed.

There were approximately twenty-one hundred soldiers in the Palmach when Shoula and Yaffa joined in 1946. Though the Haganah itself trained women only for static defence against Arab attacks, the Palmach accepted women as equal combatants. By 1947 sixteen percent of its soldiers were said to be women. It was the only organized Jewish military body, and essentially it functioned as an autonomous army, though under the Haganah umbrella. In almost all ways women were trained and expected to perform in the same capacities as men.

Shoula was a fair-haired nineteen-year-old, and Yaffa, one year younger, a vivacious redhead. They were assigned to a unit that consisted of about thirty people, of which half were female. "It wasn't unusual for girls to join," says Shoula. "It was very natural for us. We trained with the boys and never considered anything like 'women's lib.'"

In addition to their training, they helped work on the kibbutz, living in tents, which would be hastily taken down whenever there was rumour of British inspections. Then they would merge with the people of the settlement.

Yaffa: "Our group took a vote and decided that boys and girls could share the same tent. We either had two girls and a boy, or the reverse. We were so shy and naïve that there was no problem with co-habitation,'' she smiles, ''though that might be hard to understand today.''

The days began early with long hikes, running and jogging. ''They really stressed physical fitness,'' Yaffa continues. ''We had to run three kilometres each morning in hiking shoes with knapsacks on our backs. We had to know how to run long distances in the field. We also learned judo, first aid and weaponry. We learned to shoot and throw grenades. They even taught us to fight with sticks, because it was illegal for us to possess weapons under the British Mandate.''

Of immediate priority were operations to smuggle in the survivors of the Holocaust. The Haganah raised money to buy old ships, anything that could float, got people out of the displaced persons' camps and smuggled them into the country at night. Yaffa: ''The Palmach waited on the shore for them with our trucks. They used to get stuck in the sand, and we would go down into the water, get the people off the boats, sometimes carrying them on our backs, and load them in the trucks. Then we would rush to the kibbutz. There they would be mixed with the residents. When the British arrived to inspect the settlement, the refugees had already changed their clothes and lined up with everybody else. The British would ask for everybody's name, and no one would answer. They could not find out who was a member of the kibbutz and who had just arrived. We had powerful emotions when we saw the Holocaust victims. Anger and fear. Anger because something like that could happen and the world remain silent. Fear that this could happen again to others and maybe to us.''

Shoula: ''Another main objective was to procure arms. Since it was illegal to carry weapons they assigned the task of smuggling guns to the girls. We would hide them under our clothes. One time I had to smuggle some from Tel Aviv to the kibbutz. The bus I was on was stopped at the station and British soldiers came on to check the passengers. I was trying to think calmly, but I had butterflies in my stomach. If you were caught with weapons they would shoot you. The soldier walked past me. I guess I didn't look suspicious. I was really lucky I didn't get checked.'' The Palmach counted on the gentleman's ethics of their adversary to keep their gun-running women safe.

By 1947 the British wanted out of Palestine and a solution to the increasingly widespread violence between Arabs and Jews and against themselves. The Palmach, though not primarily offensive at this time, did launch attacks on British military installations at night — mainly to steal equipment. The Etzel and the Lehi continued attacks on individuals. Furthermore, the British were being pressured by Washington to allow a hundred thousand Holocaust survivors into Palestine, and pressured by the Arabs not to break the immigration ceiling. There were a hundred thousand British or Commonwealth soldiers in Palestine in 1945-46. As they came under personal attack by Jewish organizations their conduct understandably altered.

A young teenage student living in the Talpiot suburb of Jerusalem remembers that by 1946 Jerusalem had become "a city of fear. The British soldiers policed the streets drunk and wild," says Shifra. "But despite the fact that I hated them for stabbing us in the back politically, I felt sorry for the soldiers who walked around in mortal fear of being assassinated." She disagreed entirely with the tactics of the Etzel and Lehi and says, "They were ragtag, disorganized and unethical groups. They gave us bad reputations, and because of them we lost some sympathy abroad. I almost got arrested because of their actions."

Shifra was the only child of Orthodox parents. Her mother had died when she was very young, and she was raised by her father, a wealthy businessman who had worked with the British in the Second World War. She attended a French high school because her father wanted her to learn the French language and someday study abroad. In 1945 she joined the youth group of the Jerusalem Haganah. The blue-eyed, chestnut-haired girl had sensual good looks and though only sixteen seemed older. Because of the loss of her mother she had taken on adult responsibilities. She returned home from long school days to cook dinner for her father and helped him run the household. The end of the World War had produced in her "a personal consciousness of my own destiny and a sense of politics," she says. She had therefore joined the youth group without telling her father.

"My father would not have approved," she explains. "So my

girlfriend's mother covered up for me by saying I was staying at their house on weekends and holidays." As an Orthodox Jew her father was a pacifist and believed in traditional roles for women. "He would have liked not to see wars, and would ask, 'Why my little girl?' but by 1945 things were seething and boiling in Palestine and I wanted to be part of it."

Great Britain now turned to the United Nations for help in finding a solution to the conflict between Arabs and Jews that would ensue once the British had left. All Palestinians knew that there was little hope of avoiding war. As one Jewish leader said, "If the UN votes for a partition, the Arabs will not accept it. If they vote against a partition, we will not accept it." Both sides geared up for war.

In early 1946, at nineteen years of age, Hanna Armoni married one of the Lehi boys who had recruited her in 1943. They had fallen in love in part because of the shared idealism that made all members very close. They asked the organization for permission to marry, not because they needed to but out of respect. Though they were told not to expect any special treatment such as the opportunity to go on joint missions, she says, "We did seem to get them quite often."

Hanna didn't know she was pregnant when the Lehi planned their next action, an attack on the train repair factories near Haifa. Over forty members took part. Hanna was assigned with a man to mine one section of roads leading to the factories, to act as a road block. Other groups blocked the other routes, and the rest were to sneak into the huge buildings and blow them up.

Hanna and her colleague took lamps to light up the warning signs they placed before the mined section of the road in the pitch black night. It was an extremely dangerous area because there were British military bases and police stations nearby. The group sent in to blow up the buildings had twenty minutes to complete their mission. Hanna's husband was among them.

When the explosion was set off the groups began their hasty retreat, but two members did not turn up at the prearranged spot. The others went back into the huge complex to look for them. When they finally found them they rushed to leave the area, but it had been forty-five minutes since they had begun the action and the British troops in the area had already cleared one road of its mines. The Lehi group was ambushed as it retreated. Eleven Lehi members were killed and twenty-three captured; only a few managed to escape. "My husband was among the nineteen boys who were captured and all were sentenced to death," says Hanna. "Four girls were also caught and sentenced to life imprisonment."

"Those eleven people killed were our close friends. This left a lot of scars. I kept thinking over and over again about how this action had gone wrong and what we could have done to prevent the tragedy. But our main problem at this point was to stop the British from executing my husband and the others. We put up posters and sent messages throughout the city that said, 'If even one of our people is killed, we will cover the British with blood.'" The sentences were commuted to life imprisonment.

"My husband and I smuggled each other letters because letters were permitted only rarely and those that got in were censored. When I found out I was pregnant I smuggled in the news. He wanted to know the exact day his child was born, so we arranged to put an announcement in the newspaper that Sarah and Elimelach had a boy/girl, and the time." Hanna gave birth to a daughter, and her husband read the news the same day.

"All the prisoners made cards for my daughter's birth. I still have them. My husband wrote a poem for her inspired by the tragedy of a ship full of refugees that sank after the British had forced it back to sea. That moved him so much that he wrote the poem and dedicated it to our infant daughter. I think he believed that when she grew up and read it, she would understand his beliefs and motives for the things he had done."

After the birth of her daughter, Hanna returned to live with her parents in Tel Aviv. Six weeks later the Etzel and Lehi broke into the Acco Prison to get their people out. During the escape British soldiers shot and killed Hanna's husband, who was in a get-away car.

"I had often thought about the possibilities of him or myself dying," she reflects. "But it was a great shock. He was only twenty-two, but I believe that in his short life he accomplished more than most people could in a hundred years. It took years to understand, to get over the sorrow, but now it has its proper place.

"Age was an important factor. I think that if such a thing happened today, God forbid, I would be much more frightened, much less strong. But then I had strength. I told myself there would be better times, when we would no longer need war and could live in peace.

"Still, I had my baby and my battle to fight. I refused to stop. I went to Haifa and one of the member's wives took care of my baby. I kept very busy. I was made responsible for education of the youth groups."

Despite the united resistance agreed upon by the Haganah, Etzel and Lehi immediately after the end of the World War, the tactics utilized by the Etzel and Lehi, predominantly acts of urban terrorism, were condemned by Jewish leader David Ben-Gurion. On 22 July 1946 the Etzel's assault group blew up Jerusalem's King David Hotel, the British military headquarters. Eighty-eight people were killed, including Arabs and Jews. The Haganah immediately denied involvement in the action. The united resistance had come to an end.

Shoula and Yaffa remember the split. Shoula: "They did not have the same philosophy as us. The Palmach believed in building the state slowly by settling the land and protecting the settlements from attack. The Urgun and Lehi thought the way we worked was too peaceful." Yaffa: "We acknowledged their devotion and bravery but took those things for granted, as everybody was ready to give all he or she had. We did not agree with the tactics they used. We felt they were irresponsible and hurt our cause."

The two women continued living on the kibbutz. Their operations were only preparation for defence against future attack. On one exercise they spent nine days in the desert. Yaffa: "We hiked through the desert and did not see a soul. Everybody carried what he needed on his back — food, blanket, everything. We found water in watering holes. We could only drink one gulp of water every hour." Shoula: "It was a test to see who could resist drinking the water, except when absolutely necessary. We went from the north of the Dead Sea to the south."

"Now there is a road there," laughs Yaffa. "But then it was just desert. The girls did not get any special privileges. Each girl had to carry her own things too."

Both women spoke of the close friendships that developed among the group. Shoula remembers even stronger feelings for their commander. "I started to fall in love with him," she laughs. "The thing was, he was in such a high position, he gave orders and told us what to do. We were so young that I found that very impressive and fell for him!"

"I knew all about it," says Yaffa. Eventually Shoula and her commander became involved, but they separated when he was posted elsewhere. Years later she ran into him again. "But it didn't work for me anymore," she says. "I no longer saw him as this big, powerful person. Still, it was very nice to see him again."

When their training period ended Yaffa and Shoula were separated. Yaffa took a course to become a Morse code and communications officer and was eventually posted to a unit in Jerusalem. Shoula remained on the kibbutz and trained new Palmach members, including Yaffa's younger sister.

Early in 1947 a UN commission visited the Middle East to examine alternative solutions to the growing conflict in Palestine and find a viable way of moving towards withdrawal of British forces. The majority of the commission eventually recommended the partitioning of Palestine into two states, one Arab and one Jewish, with the holy city of Jerusalem administered by the United Nations. Though many Jews felt partition was a concession, the majority welcomed any suggestion of a sovereign state. Arab leaders denounced the partition plan.

The United Nations General Assembly put the motion to a vote on 29 November 1947. After much last-minute bargaining and pressuring that weekend, the vote for partition was passed.

As Ben-Gurion announced the birth of the independent state of Israel that night in Jerusalem, the first Arab attack was launched and unofficial guerrilla war began.

Yaffa: "I was in Jerusalem that night. We had gone there to see the celebrations in the streets, and we were walking in the street when we heard about the first Jewish casualty. A man had been killed on a bus. That was the first death. Arabs began attacking the roads, shooting at buses. This developed into something very serious. We could hardly travel on the roads. Thirty of our friends from the Palmach were killed this way before the "official" war even started. When the war started I didn't have a brigade anymore, because everyone except about thirty or forty people was either dead or wounded."

Shifra remembers the night of 29 November 1947 clearly as well. "That same night the fighting started," she says. "Talpiot, in the north of Jerusalem, was particularly vulnerable. It was located very near Arab villages, and the roads into the city were cut off. We had always lived very nicely among the Arabs. We bought produce from them; they came to our home. But suddenly the Jewish civilian population of Talpiot was told to evacuate. It was too dangerous and too difficult to protect. My father didn't go, because I refused to go. I stayed in our old stone house. It was next door to the home of Major Cunningham, Chief of Staff of the British forces in Palestine. Only a few doors away, in one of our most beautiful homes in Talpiot, the Haganah set up headquarters once the owners left for the city."

Because of the British chief of staff's residence, Shifra's street was patrolled night and day by British soldiers. Residents of the street were taught a password that got them past the road-block. "I remember arriving home late one night and they stopped me and asked me for the password. They were about to shoot me because I could not remember it. Luckily I spoke English and explained where I lived. When I got home my father was screaming at me, he was so upset."

Shifra's cousins had all left Palestine to study abroad. It is what her father had wished for her. She says, "It is hard to explain the kind of outlook we had then. It was completely different. For young people the key things were discipline and obedience. Especially in the Haganah." But despite the duty of obedience to her father, Shifra made the decision to stay in Palestine and fight. "I had become pretty rebellious by then. I was determined. But still, I didn't want to hurt my father so I tried to talk to him, to explain my feelings." Shifra joined the other Haganah women in Talpiot in helping to staff the telephone exchange. "The boys were already on guard in the trenches we had dug in the area,'' she says. The Haganah pressed every able person into the fight.

Hanna Armoni was also in Jerusalem when the UN vote was declared. She had moved there to continue work with the youth groups of the Lehi, leaving her baby in the care of friends. ''The Arabs started rioting and killing people and blockading the roads,'' she remembers. ''According to the United Nations we did not possess Jerusalem, but we didn't want to lose it. It is our most important city.'' Jerusalem, holy to Jews, Christians and Moslems, immediately became embroiled in war. ''The Lehi decided to stay in Jerusalem and fight to keep it from falling into Arab hands.

''I was staying in a residential area on Ben Yehuda Street, where I had rented a room in an apartment house of four or five storeys. Some of our members had escaped from the Jerusalem jail, and one came to stay with me and my friend.

''One morning I woke up very early and began to read a book. Then I heard some shots and went out on the terrace to look. I saw smoke coming from a car, and I knew exactly what it meant. The British had put two cars full of explosives on each corner of the street. We were on the fourth floor. I woke up my friend; then we rushed to try and wake up the others. All this was just a matter of seconds.

''I didn't hear a blast, but I remember a lot of dust, and the terrace falling inside the room. I think I lost consciousness for a moment but when I came to I wondered, 'Am I alive?' Then I said, 'Cogito, ergo sum.' There were bodies everywhere. It was so awful. I had never seen so much blood before.

''Our house was the only one that remained standing. The others were just gone. The British blew up houses in this part of town three times. They wanted revenge against the whole Jewish community, I think.'' The terrorist tactics against individuals in the 100,000-strong British military force in Palestine had led many British soldiers and administrators to sympathize with the Arabs. Fear and hatred sometimes led to acts of revenge by the British. The incident described by Hanna fits the pattern of several private acts of vengeance by the British against Jews: ''Some of the Lehi stayed in Jerusalem, but most enlisted in the Jewish army, the Zahal, which had formerly been the Haganah." Though previously against employing women in combat, the new army, in its sudden move to expand and transform itself into a regular army, took anyone, male or female, who could fight. Women with children, however, were out of the question. Hanna says, "I went to enlist but they wouldn't take me because I had a child. I felt so empty. It was a very hard period. I had been fighting for so long, I had almost been killed so many times, but when the final war started they wouldn't let me fight."

The UN resolution provided a six-month timetable for British withdrawal from Palestine, with the final evacuation scheduled for May 1948. Jewish leader Ben-Gurion ordered his people to begin procuring arms for the war that both sides knew was inevitable. Because Israel would not officially become a sovereign state until May 1948, Jews were denied legal importing of arms and the British continued their naval blockade. Furthermore, Jews were still not permitted to own weapons. The Zahal organized clandestine shipments of salvaged World War II equipment, purchased illegally abroad with financial help from Jewish communities in several other countries. The Zahal also welcomed volunteers from abroad, particularly men with military expertise.

When the partition was announced both Palestinian Arabs and Jews had pitifully small forces, which were illegal in any case under the British Mandate. At the beginning of 1948 it is said that the Palestinian Arabs had only 2,500 riflemen. The Arab liberation army that was quickly assembled after the UN vote consisted of 3,830 volunteers.

Guerrilla tactics were utilized by both sides. Both had small defence forces to guard and protect their settlements and villages, and both also struck out offensively. Much of the fighting took place at night, hit-and-run affairs and road ambushes. The Zionist forces of both the Irgun and the Zahal attacked a number of Arab villages — both inside and outside the area designated Jewish by the United Nations. The forces went in with horns blasting, recordings of Arab wailing and screaming, warning Arabs to get out before the blood-bath. Arab guerrilla units struck out at isolated kibbutzim and transport convoys.

Armoured cars were a necessity as road ambushes became

increasingly frequent, particularly in isolated parts of the country. The Zahal made its own by cementing steel plates to commercial vehicles.

Yaffa worked at a kibbutz near Jerusalem, functioning as "a secretary without papers, meaning I did everything no one else had time to do." She adds, "Everybody else had a specific thing to do. One would be in charge of arms, which were hidden in deep holes in the ground. Another was a driver. Another was the officer. I feel funny saying officer," she smiles, "because he didn't wear a uniform or anything. He didn't look special. He was a friend in every way. We didn't have special dining rooms or anything. We all ate together. I was in charge of communications. I was the only woman there, but we had people to cook for us, so I didn't have to do stereotypical female jobs. In the headquarters I shared a room with one other guy, but nothing went on between us. If I wanted to get dressed or undressed, I would just say 'curfew' and he would turn his head. Curfew meant the other one had to hide under the sheets or turn around. I felt secure with him in the room even when I was undressed. I never worried about it. A lot of people can't understand that. I lived for six months that way and nothing ever happened, because we weren't in love. There was no reason for us to get into any other relationship than the true friendship we felt for each other."

Yaffa came under fire for the first time when a convoy she was in en route back to Jerusalem was ambushed. "As a Morse code operator I was really supposed to be stationed behind the lines, but there were no real lines. My commander and I were asked to lead a convoy to a certain area, so we were in the first truck. The convoy was attacked and my commander was shot very near his heart . . . about an inch away. I was fine."

During this time all limited resources available to the Jews had been used to fortify their settlements, most of which were in the coastal plains, Galilee and the northern Negev, in hostile Arab territory. The roads were supposedly under British control, but Arab ambushes made transportation in and out of each kibbutz difficult and dangerous. The Haganah organized convoys with their home-made armoured vehicles, guarded the trucks bringing in food and supplies and helped evacuate civilians from kibbutzim in the most dangerous, isolated areas. Shoula was now stationed at a kibbutz in the Negev, where she worked in the armoured vehicles in the convoys. "There were holes in the cars, where we put our Sten semi-automatic guns. We prepared to open fire at any minute. Many of the convoys I travelled with came under ambush. Sometimes we would get an SOS signal over the radio that a certain settlement was under attack. Then we would go there to try to get the women and children out."

Fighting in Jerusalem continued day and night. Shifra remained in her old stone home in Talpiot. Her father helped get food and water supplies to the beleaguered forces but eventually left the suburb and stayed in the city.

Talpiot was on the road from Bethlehem and was surrounded by Arab villages. It had become a military zone. Shifra was given a gun. The tumultuous and quickly paced events have merged and become confused in her memory. She says, "It is all fused. I guess I've blocked some of it out. I remember walking around with a gun all the time." Her unit was given orders to capture the rich Arab neighbourhood of Baka, a first step towards opening the road into the city. "When we got this order everyone was extremely afraid," she remembers. "We had to go through the Arab neighbourhood in units of three people. We planned the attack for night, as we always did. We were to go in and take the area, house by house. None of us thought we would leave there alive.

"We walked in and it was incredible! The coffee was still hot and the beds had been slept in, the closets were full of food and clothes. The entire population had fled. We just walked in and took it.

"Some of the houses had already been looted — by our own people. There are many things that we are not proud of."

Also merged in her memory are the endless nights spent in trenches. "The Arabs attacked by night, so we had our guard shifts. We would just lie there waiting for a sound. We sat there and didn't know if they were coming or not. We didn't hesitate if we heard anything. We began shooting. I killed people.

"I never allowed myself to think in terms of where I got the courage. We had an attitude: it was live and let live, but the

Arabs weren't going to let us live, so we had to fend for ourselves. It meant I had to fight. We took care, tried to be careful, but we didn't think in terms of courage. We just did it. Maybe there were times I became hysterical and someone had to slap me out of it. I don't remember. But I never saw the people I killed. I never shot point-blank except at target practice."

Jerusalem was under siege for six months, from January to June 1948. The holiest city was the scene of the bloodiest war since the Crusades. Hanna, having been denied a chance to join the fighters, had returned to her baby and parents in Tel Aviv. Yaffa remained with her unit in the city. She remembers being given a small pistol, but it was "more for myself than to kill them — to show my courage and if necessary to kill myself."

Mortar shellings went on ceaselessly, day and night, throughout the city. There was very little food left and most of the water supply was gone. What little remained was distributed to the hospitals, now overflowing with wounded.

Yaffa: "You asked me if we ever broke. In a way I did. When the fighting really began, my officer would lead our fighters every night. I used to wait for them to get back and wonder who would come back alive this time. During the day I was in charge while the commander slept. One of the things I had to do was to attend funerals. Every day there was one. These were our friends. I had to go to represent the officers, to show that those who died were not ignored or forgotten, to give courage so our people would not think that if they got killed it meant nothing and we would forget about them. I made a point of not crying at the burials. I acted strong.

"Then one day somebody very, very close to me got killed. I didn't go to the funeral. It was my duty, but I didn't go. I knew I couldn't stand there without crying. I found an excuse not to go. I stopped going to funerals after that. It was a kind of breaking point for me."

Shifra: "All those months, people I loved, friends of my youth who were fighting on all fronts in Jerusalem were being killed or maimed. Some were taken prisoner at a massacre at the Etzion settlement between Jerusalem and Hebron.

"One morning, coming off duty, about ten feet away I saw one of our armoured vehicles with four of our friends lying in it with their guts and brains spilled all over. Dead. I was physically sick. I have never forgotten seeing a friend of mine with her head blown off."

The seventeen-year-old girl was changing. "I was becoming more resilient, tough. And also cynical. Maybe cynical is too strong a word — maybe just weary." She met a British Jewish soldier who had left England to volunteer his expertise to the Zahal and help organize medical care. He would become her husband, and she credits him with "saving me from becoming too far gone. . . . He helped me to believe there was hope in the world. He said someday everybody would be able to live in their own country, under their own fig tree. He helped me a great deal."

Shifra's unit fought in Operation Nahshon, an attack to reopen the road into Jerusalem from the north. But as the Haganah fighters fought to reclaim mile after mile of the road that led in from the heights, the IZL (Fighters of the Irgun) attacked a nearby Arab village of Deir Yassin, killing two hundred men, women and children. Shifra witnessed the results of the massacre only hours later. "When I saw what our side had done I felt sick to my stomach. I felt apologetic. But I shouldn't have because this was the Irgun — Menachem Begin's doing — not ours. He was smearing our war with shame." The tragedies and atrocities of the unofficial war did not spare innocent civilians. Guerrilla fighting obstructed order and front lines. Civilians huddled in the basements of both mosques and synagogues.

But in this "total war" for each block of each city, each acre of every settlement, only the Jews used women as fighters. There may have been exceptional cases in which Arab women picked up guns, but for the most part Arab women would not fight until later conflicts. It was something they learned from the Jews. Arab reaction to enemy women fighters was reportedly demoralization. More than affronting their chauvinism, the presence of Jewish women in battle clearly indicated to the Arabs that this was a war in which the Jews were prepared to use every possible force.

Though even the Palmach tried to keep their women fighters relegated to the jobs of arming convoys and defending settlements — to protect them from the most severe combat — the women fought alongside the men and suffered injury, maiming and death. Most frightening were reports of what happened to women soldiers captured by the enemy. "Our commanders did feel protective," says Shifra. "I am sure many of them would rather have made love to me than send me out to fight. But they had no choice. It was total war."

A founder of the Palmach, Yigal Allon, later wrote about the importance of women in the fight:

> The presence of women in combat units blurred and decreased the harshness of military life; it lent substance to the Palmach concept of an armed force free of militarism; and it precluded the brutalization of young men thrown into an all-male society for months on end. The mobilization of daughters, sisters, sweethearts, and often wives turned the Palmach into a true People's Army.[2]

Shoula agrees: "When a woman was in the group the men behaved much more nicely and politely. They seemed happy to have us. I don't think we affected them in any kind of negative way. Sometimes, I thought, women were even stronger emotionally in bad situations. In war some men are strong and some are not. It really boils down to education. If you are educated and trained to do your duty, you do it, whether you are a man or a woman. Women don't like killing, perhaps — but neither do a lot of men. We felt terrible when anyone was killed or maimed, and did not take it any more badly if it were a girl rather than a boy. We were in it together."

Yaffa was permitted a short leave to visit her family in Haifa in April 1948. Exhausted from the past four months in Jerusalem, she welcomed the break. "I arrived home on April twenty-first," she recalls. "After talking to my parents for a few hours, I went to bed and fell asleep. I was so tired."

That same day the British commander, General Sir Hugh Stockwell, announced that final evacuation of remaining British troops in Haifa would be completed immediately. With the British only hours from departure, both sides prepared to battle for the city that was almost exactly half Arab, half Jewish. The Arab mufti of Haifa instructed all Arab civilians to leave their homes until his army had claimed the city.

The fighting began that night. Seventy thousand Arab civilians fled their homes as they came under fire. Most of them left by sea and landed in Lebanon and Syria, becoming the first Palestinian refugees.

Fighting was intense. The Jews had the strategic advantage of living on the heights, and they threw down flaming barrels on Arab strongholds below.

In the morning it was all over. Yaffa says, "My parents came into my bedroom and woke me up. They said, 'Haifa is ours.' I couldn't believe it — I had slept through the whole battle! I hadn't heard a thing."

Shifra had a similar experience in Talpiot, Jerusalem. She was still in her father's old home, which by now had been robbed of most of its finest treasures. "I had been on duty for twenty-four or twenty-six hours, and I had gone back to my house to sleep. I woke up to a terrible pounding on the front door. Two of our people were standing there when I opened it. They asked me if I was all right. I said, 'Of course I'm all right. Why did you wake me? I'm so tired.' They said, 'Your house was bombed.' A mortar shell had come through the wall, ricocheted and gone out the other wall of my bedroom. I didn't hear a thing! They had seen the damage from the outside and thought I must be dead."

Shifra remembers learning to get through her once peaceful neighbourhood by running from one doorway to another, dodging shells and bullets. "If you heard the sound of the shell, you were safe," she says. "It had bypassed you. If not, you were dead." Asked what her most memorable day had been, she recalls "the day I visited the hospital and saw a pile of food on the tables. We hadn't seen food for so long."

Just before the final withdrawal of the British from Palestine Ben-Gurion sent a personal envoy on a peace mission to Amman to discuss with King Abdullah of Transjordan means of avoiding full-scale war. His envoy was a woman who would be famous to the world in years to come. Disguised as an Arab, Golda Meir made safe passage to the top-secret meeting. But the talks broke down when she was told that the Arab nations would call off a full-scale attack only if the Palestinian Jews would relinquish claims to a sovereign state. Mrs Meir, who would lead her country to war in years to follow, relayed this final ultimatum to Ben-Gurion. A few weeks later, only two hours before the official end of the British Mandate on 14 May 1948, Ben-Gurion read Israel's Declaration of Independence to an emotional audience in Tel Aviv. Hanna Armoni, then residing there with her family, remembers that the first Egyptian air attack on the city came that same night.

The next day the new state of Israel was simultaneously invaded by armies from Lebanon, Syria, Egypt and Transjordan. In total the invading armies were said to number about twenty-three thousand men. The Arab armies had tanks, superior firepower and air support. The Haganah, which overnight had become Israel's official army (the Zahal) had no tanks and no air force.

Fierce fighting continued in Jerusalem, where Zahal fighters attacked and attempted to seize the Old City as soon as the British were gone. In the month-long fighting that ensued the Jewish quarter of the Old City was almost completely levelled, and the Jewish fighters eventually surrendered to officers of the Arab Legion. In other sections of the city water and food shortages had become critical. On 11 June a truce was negotiated by the United Nations, and fighting throughout the country temporarily stopped. Shifra was given five days' leave and went to stay with a friend near the beach in Tel Aviv. She says, "I just stayed on the sand, staring at the water for the whole five days. I was burnt to a crisp, but I didn't care. I hadn't seen water for so long. I kept staring at it, and it helped me a lot. It helped me get back to myself."

Both sides used the month-long cease-fire to regroup and resupply. The Israeli army now numbered nearly sixty thousand soldiers, of whom twelve thousand were women, and was in the process of becoming a conventional army. As its numbers grew the policies regarding women changed. Because it was now the official army, run by the government, political as well as practical arguments favoured taking women out of direct combat and putting them in their own units. As a large body of the electorate, Orthodox Jews pressured the leaders to stop women from fighting. They and others argued that it was immoral to billet men and women together; some statistics indicated that mixed-combat units had higher casualties; no other civilized country accepted women into combat roles in the armed forces. Because the Israeli army no longer needed "every able-bodied man or woman," and because pressure was extreme, the army enforced the change. Where possible women were assigned new roles, out of combat.

Shifra was sent to an officers' training course in Tel Aviv. "They taught us to forget everything we had been previously taught and learn things all over again," she muses. "It meant a new kind of discipline, and even uniforms. I actually preferred it. I liked the change because I am a very organized person. I got a rank and became a sergeant major in charge of a nurses' unit." But was there no resentment, after having been a fighter? "No. Women continued to do vital jobs that allowed the men to go forth and fight. These jobs had to be done by someone, and I don't think they were any less valuable to the war effort. There was a fear of what might be done to a woman caught by the Arabs. And anyway, we girls weren't fighting to prove anything. I have no hang-ups about my femininity. I fought when there was no choice, but when I was no longer needed in that role I was happy to move on to something else. Besides," she smiles, "I had my uniform tailor-made and looked good in it."

For Yaffa and Shoula the change had no personal effect. Though the Palmach was no longer sending new women recruits to mixed fighting units, they did not reassign those who were there. Shoula remained in the Negev where fighting was intense and continued in the same capacity. Yaffa remained a communications operator in a fighting unit, and her commander's adjutant. When asked if they nonetheless resented the policy change and its implications, they both seemed surprised. Sexism was hardly an issue while the fighting lasted, they said. Yigal Allon wrote about some of the reactions:

> The girls stormed at any proposed discrimination, arguing that it ran counter to the spirit of the new society being built in Palestine to restrict women to domestic chores, particularly since they had proven their competence as marksmen and

sappers. In the end, the wiser counsel prevailed. The girls were still trained for combat, but placed in units of their own, so that they would not compete physically with men. Whenever possible, they were trained for defensive warfare only.[3]

Yaffa says, "They stopped letting girls go in action units where they had to run a lot. Physically girls couldn't run as well — and if they couldn't run fast enough they could endanger the whole unit, so they were put in other units. But they still fought."

Shoula: "Yes, I remember they made decisions on these things. They found it very hard for women. But mainly I think it was because of political pressure."

On 11 July fighting broke out again. The Israeli forces in Jerusalem concentrated on lifting the siege of the city. In the Negev, Egyptian armies driving tanks across the sand struck out at isolated Jewish settlements now fortified by Israeli defences. The fighting lasted ten days before another cease-fire was arranged. Over the eight-month war, a total of four cease-fires would give the opposing forces respite and a chance to feed their civilians.

Shoula was stationed on a settlement that was now behind Egyptian lines. The kibbutz had been cleared of all its women and children and was fortified with barbed wire. Trenches had been dug in the gardens, and air attacks had destroyed some of the buildings and shattered most of the windows. Despite the tragedies, she remembers that the soldiers did their best to keep spirits high. When their chicken coop was bombed they organized a barbecue, and Shoula and a friend tore down curtains in the abandoned nursery to make themselves dresses. "Even in all that," she laughs, "we wanted to look pretty for the barbecue." But for the most part the women lived in khaki shorts, with only the barest essentials for cosmetics. "I didn't even have a mirror," she sighs. "We tried to wash our face, comb our hair. We were so exhausted that we slept marvellously. We were tanned and fit, and even without the clothes and make-up, many of the girls looked great." A wartime photo of Shoula shows a glowing beauty, despite the toll taken day by day as the tragedies of the war continued.

When a settlement at Bruchayim came under attack, Shoula and a female friend volunteered to work on the convoys that were sent to help evacuate those who could not fight. En route the convoy was ambushed. "My girlfriend, who was riding in the same car as me, was shot through the intestines. I hadn't realized, and rushed with my gun out of the car to fight the attackers. But somebody called me back and told me my girlfriend was calling for me. I went into the car and she begged me to stay with her, not to leave her. She was conscious only until we got her to our hospital in Niram. I stayed in the hospital until they had finished her surgery. It took them seven hours to sew up her intestines. She was so badly injured that I thought her family should be told. I hitch-hiked a ride in a one-engine plane and went to her family and told them. Her father took it very badly. I told them they couldn't come to this place, because the Negev was cut off by the Egyptians and was very difficult to get to. But her brother, who was also a soldier, managed to get there somehow. Some days after this I saw him come through the door of the hospital room. He saw her lying there and smelled the terrible smell, and he fainted. I stayed with her for three weeks, until she was able to get to a good hospital. It is a miracle that she lived."

Shoula herself was wounded slightly in a convoy only weeks later. On a five-kilometre journey along mined roads the truck hit a mine. "Suddenly I found myself flying through the air," she marvels. "I hadn't felt the truck hit the mine, but just the sensation of being blown up, thrown through the air and falling very badly. All the people were thrown out of the truck except one girl. I remember her crying. She had lost her kneecap. I had broken my leg."

When she was able, Shoula returned to the kibbutz, where she sometimes took a break from the convoys and did other jobs, including working with the Morse code operators. She also continued to train recruits. She had become an officer in the Israeli army and says she never had any trouble training the young men. "It's all a matter of education," she smiles. "If kids from kindergarten are taught that women should be barefoot and pregnant in the kitchen, that is how they will treat them. But our kids were taught that both sexes were expected to fight — to do their duty. So I trained young boys and girls, and never had any problems with obedience or respect."

Yaffa and Shoula worried about each other often. They also worried about their sisters and brothers who were soldiers, stationed throughout the country, and their parents who had volunteered for other work, in Haifa. Both girls' fathers helped guard the roads near their city in a unit of the "Home Guard" for which very young and very old men could qualify. Their mothers did volunteer work in canteens or cafés for the soldiers.

Shoula's brother was killed during these months of fighting. She also lost a great many of her friends. But she never lost faith. "I felt very sorry, but I never broke down," she says quietly. "If somebody we loved was injured or died, we had to be strong. We shared the sorrow together. We always tried to keep each other's morale high. Once I was in an armoured car near Tel Aviv, and we got the news that thirty-six of our students had been killed near Jerusalem. We felt so terrible, but we were together. We were like children, so young, and we clung to each other for emotional support. I became very emotional when I heard this news. But a friend said, 'Be strong, because we need the strength to tell another girl that her brother was there.' He told me to be strong for someone else."

Yaffa's unit was moved around from one post to another as the situation required. She remembers worrying about Shoula, fighting in the Negev: "The Egyptians had almost reached Tel Aviv — they were very deep into Israel," she recalls. "I was stationed in the city of Jaffa where I was a Morse code operator. There were degrees of urgency. There was a regular message, there was a more urgent message and then there was BAHOOL, meaning 'Absolutely critical — it can't wait.' I listened to the bases in the Negev sending these urgent messages. They were calling for help because they were under attack. I was so scared for Shoula, and I was feeling the weight of these responsibilities. Here I was, not even twenty years old, and I was trusted to get these critical messages to the right people. I was trusted with such a great responsibility."

Shoula was at that moment experiencing an air attack. She remembers, "We had gone outside to the outhouse, to clean it. We were carrying pails and everything, myself and another girl. We volunteered to do it at a quiet moment because during wartime these things get neglected.

"Suddenly we heard the airplanes. We ran out of the outhouse and jumped into some trenches we had dug outside. I was looking up, and I saw the plane and the bombs falling down. From that perspective, it appeared as though they were going to land right on top of us. So I said a Jewish prayer that you pray before you die. I really believed this was it."

The bombing stopped and the young women were safe. "I had never felt like that before," she says. "My knees were so weak. At other times, many other times, we had been bombed. There were times when there was shooting like crazy from their tanks and guns, and other times we were bombed. But we had a cover. Something over your head makes you feel safe somehow. But this time it had been in the open. It was really frightening.

"Normally we didn't think about death very much, only when they brought in our friends without legs or missing a hand or without eyes. We felt the sorrow of death, but never believed it could happen to us."

Hanna Armoni spent the eight months of the War of Independence at her family's home in Tel Aviv. She says, "I never felt more empty in my life." In the final irony, the young woman who had committed herself to fight for Zionism, by any means, was denied a role in the war that would finally win it. But tragedy was not over for Hanna: her mother was killed in an Egyptian bombing raid of the city. "I tried to be strong." Reflecting on the death of her husband and the events of her young life, "I had lost people, but they had not gone as helpless victims. My husband's death had not been in vain. He died a fighter. So even if I missed him, I knew his death was not for nothing."

In December 1948 the Israeli army finally took the initiative and managed to break through Egyptian lines in the deserts and encircle the Egyptian army from Gaza to El Arish. The fighting was intense, but finally the Egyptian government called for an armistice to end the war. On 7 January 1949 the fighting finally ended. The Israelis had gained slightly more than the partition had granted them, including the whole Galilee and the Negev. Jerusalem remained a divided city, but Israel had won her independence.

At least a million Arabs were displaced during the war for Palestine. They would become and remain homeless refugees. Their plight has remained a catalyst for wars that have seared the region in recent decades. In their search for a home they have put into practice some of the lessons learned from their Jewish neighbours: today the participation of the Arab woman as fighter or terrorist is common.

Hanna Armoni remarried after the war. Her second husband was also a Lehi member and a close friend of her late husband. He too had been captured and imprisoned at Acco Prison after the rail factory attack in 1946. They had two children, both daughters, and she lived with him happily for the next twenty-five years.

"The stigma of being in the Lehi lasted, though," she says. "Ben-Gurion disagreed with our tactics, and though he later said he admired our courage, for a long time no Lehi members received good jobs or government posts." Later things changed. Yitzak Shamir was a Lehi member himself.

Shifra married her English officer and managed to take the final exams that she needed to graduate from high school. They moved to a new settlement, "with very few amenities," where she gave birth to the first Jewish baby who had been born there in two thousand years.

About staying in the army, she says, "I wasn't overly ambitious to do that," then adds, "I must tell you that I have absolutely no hang-ups about feminism. I enjoy being a good wife and a good mother. I know I am intelligent, I know I am good at things. But when somebody says to my husband, 'You have a charming wife,' I don't get upset. I take it as a compliment.

"We women of Israel fought because we had to. We fought because we had no choice. It wasn't a feminist thing. It was our duty. And when the Haganah grew into a viable army, we weren't needed in the same way. That didn't bother me. I feel women are useful and necessary in other roles in our defence."

The war left its scars. Shifra, who later moved with her

husband and children to the United States, still dislikes visiting Jerusalem. "For me it is a reminder of an unhappy childhood and all the tragedies of the war," she says. "It is hard to walk those streets without being reminded of all that."

Did the war strengthen or weaken any religious convictions? "I came from an Orthodox home," she replies, "but I didn't become any more or less religious. Not having gone through the Holocaust, the European tragedies, I didn't have the same questions to ask as they did. We questioned things, of course. But we were fighting! We didn't walk like sheep to the slaughter.

"My son was born in the States, into a different life. Mommy is Mommy to him. He doesn't regard me as a fighter or an ex-fighter. I am just his mother."

At the end of the war Yaffa continued to serve in an army camp for a number of months, and says she did not have problems because there was no immediate change. Shoula was sent by the Jewish Agency to France, where she helped work with Jewish refugees from Morocco, Algiers and Tunisia. The two friends met again about a year later.

"We were both young and single and had a lot of fun together," says Shoula.

Thirty-five years later Yaffa, who now lives in the United States, sometimes looks at her teenage children and wonders how she did the things she did at their age. "In America, they are still babies at eighteen," she says. "We were grown up. We had such responsibilities at that age. And they have nothing."

Shoula chides, smiling, "Let them be happy like this." Yaffa agrees, and continues, "You know, for a long time I didn't talk about those days because I didn't know how to say it right, and it was so important to me. It was the most important time of my life. I was so proud to have done it. Thinking back, I really do appreciate the deep friendships and the wonderful ideological life that we shared."

Shoula remained in Israel, married and had children. Thinking back she says, "Sometimes I look back at those days and think how miraculous it was — what we did, and what we did it for. It seemed so wonderful, even during the war. We fought for our future."

Now women are not allowed combat roles, but Shoula thinks current wars would be too difficult for women. "But my daughter, like my son, does her duty, and if there was ever total war again, I'm sure that she and all the girls would fight. Our children don't ask questions. We have taught them from childhood that the history of our nation, from its very birth, has meant war. They know this. They know not even to ask questions. Maybe it is wrong not to ask questions, but my children know their duty."

AN EAGLE AND AN ANGEL
Indochina, 1946-54

I went to get a set of fatigues. I took a size 2 . . . because I was actually average size for the Vietnamese. When I had them and my helmet on . . . I heard men beside me whispering, "I think it's a girl. No, it can't be. No girls here. Impossible!"

Brigitte Friang,
Combat reporter, Indochina

One night we watched from the tunnel and saw all these white chutes in the sky. It gave us hope . . . to know that we weren't forgotten. . . . One young soldier landed near me, and rising to his feet he said, "Well, a woman here! I guess this isn't hell after all."

Geneviève de Galard,
Nurse, Siege of Dien Bien Phu, Indochina

The stories in this chapter are in many ways unique. They are the extraordinary experiences in the French Indochina War of a French woman war correspondent and a French nurse.

I had already met Brigitte Friang and heard of her life in the French Resistance during the Second World War. As I sat and talked with her in her two-hundred-year-old stone house, filled with Far Eastern art and antiques, I became aware of how very much she loves the culture of Vietnam and how the almost twenty years of her life spent covering the war there have continued to influence her work and remain in her thoughts. Today she is writing a novel set in Vietnam. She has already written two fascinating books about the country and the war. The first relates her experiences there as a war correspondent from 1951 to 1970: *Regarde-toi qui meurs. 2, La guerre n'a pas de fin.*

The other, her account of the fall of Saigon in 1975, is entitled *La mousson de la liberté*.

As she talked about war and her experiences in it, anecdotes from the Second World War and over twenty years in Vietnam (as well as war coverage in the Middle East) sometimes merged. She had been a participant, then a participant-observer and finally an observer of war. It seemed like a hell of a lot of war for anyone to have experienced. I was amazed that after internment in Ravensbruck and Zwodau, this petite, cultured woman would choose to go to Indochina. She laughed at my comments. "That is true, but I suppose I missed the war and the fraternity that people who have not experienced it will never understand."

Geneviève de Galard de Heaulme is a charming woman who today lives with her husband, a former French officer, in an elegant Paris apartment that has been in her family for decades. Everything about her is straightforward, simple and direct. A deeply religious woman with a warm sense of humour, she does not alienate the listener with her expression of her beliefs; instead, she seems the more open for sharing them. Geneviève de Galard is exactly the sort of woman I would want to have nurse and care for me if I were ever seriously ill. She is at the same time both strong and gentle.

I first heard of Geneviève in 1980 when I was working on a documentary television series, "Vietnam: The Ten Thousand Day War." She granted the producer an interview, one of the first in which she had publicly spoken about her experiences as a nurse during the French Indochina war. Geneviève was stranded at the French garrison of Dien Bien Phu and spent the duration of the epic battle there as the only woman, nursing thousands of wounded soldiers. Journalists dubbed her the "angel of the siege," and when it was over she was an international heroine.

The two women have never met but have heard of each other. Brigitte said of Geneviève, "She is marvellous." As children, both were fascinated with Indochina, and it was this interest, in part, that motivated them to go to the war that was rather ignored and often unpopular in France during its long and bloody nine years.

These two women have remarkable stories, but it is important to remember that there were many French women who served with the French Expeditionary Corps during the Indochina war. According to a Pentagon study, "They served at staff headquarters, in communications, special services, medical service, and parachute repairs. Most were assigned to Saigon or Hanoi. Some took an active role in field operations with troops or served as liaison pilots. Several won the Croix de Guerre." Brigitte and Geneviève became famous because of Indochina, but there were many other women who served and died there who remain largely unknown.

Somewhere in the far reaches of childhood memories were black-and-white images of jungle. Behind the creeping vines and thick undergrowth emerged a vast crumbling empire of stone temples. There were rubber plantations bordered by muddy rivers and banana trees.

Brigitte Friang was only seven when she saw the photographs in a Paris exhibit depicting life in France's prize colony, Indochina. The girl clung to her mother's hand, stared into the pictures and was lost in another world. "I said to myself, 'By God, I shall go there someday.'" recalls Brigitte. "And I never forgot it."

Twenty years later, in the autumn of 1951, Brigitte was boarding a DC-4 that would take her fifteen thousand miles from Paris to the land of the glossy photographs. But the images had changed, altered by six long years of war. She was going to set a precedent as the first, and for some time the only, woman war correspondent in Indochina.

The past six years had been interesting but not easy for the twenty-seven-year-old journalist. First there had been the struggle to come back to life after the living death in the concentration camp at Ravensbruck. Gradually the nightmares had become less frequent, and there was a slow readjustment to the mundane. Under President Charles de Gaulle's leadership those who had joined the Resistance were honoured and offered prestigious jobs in government and the military. Brigitte was appointed press attaché to the Minister of Culture, a former maquis fighter, André Malraux. Then de Gaulle left office, and a turbulent procession of revolving-door governments tried to deal with post-war economic problems and with another war

France was fighting in the jungles and rice fields of Indochina. Ironically, it was a war in which the French pitted traditional armies against the very guerrilla tactics they had used so well as maquis.

Despite her fascination with Indochina, Brigitte's decision to accept the job seems remarkable after all she had suffered during the Second World War. She explains: "There was my obvious fear and hatred of war. But I had to admit to myself that as much as I hated it, there were aspects which I missed — the cops and robbers we played. I missed the danger, the test, the challenge. I always liked to test myself, perhaps to see if I was equal to my male friends. Men are always talking about their experiences in war, and it gets quite boring. So I thought I'd like to go too, for the adventure, and to show them, perhaps."

The war without fronts was a conflict that French leaders had expected to win quickly, too soon forgetting the lessons of their own fight against a foreign power. Already the conflict in Indochina had dragged on for six years, its politics as confusing as the military strategies continually revised by the French command. In 1941 the French had surrendered to the Japanese their hundred-year-old colony composed of what are now Cambodia, Laos and Vietnam. A Vichy government functioned as puppet leadership while Free French Forces operated from outside the borders, conducting sabotage and small armed attacks against the Japanese in anticipation of a resumption of their colonial power. But by then war in Asia had become complex. The American OSS also fought the Japanese in Indochina, and Roosevelt let it be known that he believed colonialism had been the main catalyst for the Pacific war. His agents conducted some operations in conjunction with a small group of Vietnamese who were also operating against the Japanese. They were led by a man who had many aliases but who would become known to the world as Ho Chi Minh. Some historians argue that Ho was primarily a nationalist, others that he was a hardened Communist revolutionary.

When Japan surrendered at the end of the Second World War international agreement provided that the allied forces of China and Britain supervise the freeing of prisoners and round-up of Japanese troops. While the Chinese in the north were preoccupied with the spoils of war, Ho Chi Minh set up a government in Hanoi, declaring leadership of an independent Republic of Vietnam.

In the south the short-staffed British accepted help from Vichy troops, and soon the French had manœuvred to regain control. When the Chinese and British withdrew, neither French rule nor Ho's republic was recognized internationally. In the months that followed, riots and small armed attacks ignited into full-scale war. Ho and his followers, the Viet Minh, retreated from Hanoi and based themselves in inaccessible jungle camps. The French installed Vietnamese emperor playboy Bao Dai as governing figurehead and sent for reinforcements from home, among them many of Brigitte's friends from the Resistance.

When Brigitte's flight finally landed at Gia Lam airport, just east of Hanoi, she descended into the cool, damp air and her first jumbled and jet-lagged impressions of Vietnam. The drive into the city revealed the effects of a hundred years of colonialism; the Paris-like broad, tree-lined avenues were completely unlike the jungles and ruins she had imagined. But they were out there, where the war was.

The guarded press camp consisted of a block of two-storey yellow-washed villas, large rooms, restaurant and bar. The journalists who had heard that a woman was arriving were struck by her diminutive size. Even at twenty-seven, at only five feet two, ninety-four pounds, she cut an adolescent figure. Astonished that any woman, but particularly such a small one, should have been sent to cover the war, the men were, as Brigitte says, "somewhat surprised." She adds that "I was some rare bird."

At the press briefing the next day, a spokesman for the military informed them of optimistic new campaigns. The situation, in fact, was grim. Despite superior firepower, the French were limited in fighting an enemy that was everywhere and nowhere. They had abandoned many of their posts in the countryside and mountainous areas, the climate and terrain as much their enemy as the Viet Minh. This tiny corner of Asia had become a big stake for both sides in the escalating cold war. The Chinese revolution two years before and the outbreak of war in

Korea had quieted Washington's condemnation of colonialism and convinced U.S. leadership that this was not a war of independence but a communist revolution directed from Moscow and Peking. The United States began paying for two-thirds of France's military campaign. China and the Soviet Union countered by supplying the Viet Minh with help.

Brigitte, determined to get out into the field, remembers that "in the beginning there was some friction. The officers didn't want me along. They thought I might slow them down, endanger them, disrupt their operations. But they got used to me. First they found out about my involvement in the Resistance in the World War, so I gained their confidence and respect." Brigitte had been hired by the French Information Service in Indochina and was, in the beginning, asked only for "descriptive reports, or austere political and military studies of the Viet Minh. I didn't have to be, as did many of my companions, critical of the command." This also helped break the ice with the officers. Her first operations were with a battalion of the First Vietnamese Paratroopers in Saigon.

She laughs, describing her own militarization: "I went to get a set of fatigues. I took a size 2 — not 1, not 3, but 2, because I was actually average size for the Vietnamese. When I had them and my helmet on and was out in the field, I heard men beside me whispering, 'I think it's a girl. No, it can't be. No girls here. Impossible!'"

One of her first field operations was in November 1951. "I was with a paratroop unit sent into Hoa Binh, an important Viet Minh supply link on the bank of the Black River, not far from Hanoi. We had just taken over this valley and were being counter-attacked. All day long you could hear guns and mortars firing. One afternoon a young soldier I met on my way to the CO hut asked me if he could have the pleasure of my company for a walk, someday. I couldn't believe it," she laughs. "In the middle of fighting a major campaign, he was being the perfect gentleman."

French tactics by now largely revolved around superior air mobility. Whole paratroop units would be dropped into remote terrain on search-and-destroy missions that would become the basis of American strategy in Vietnam more than a decade later. Brigitte decided to learn to jump. "Ironically, it was something I had always wanted to do," she says. "Just before I was caught by the Gestapo in 1944, I was told I was being sent to England for a while to receive sabotage training and learn to parachute. I was so excited about learning to jump that it was a bitter disappointment to me that I got caught," she laughs.

She made two jumps before getting her training, simply following instructions given to her before she boarded the C-47. "The dispatcher, poor man, was green. I was so excited to jump after waiting so long that I didn't put the right leg first and push on it to jump. I didn't do it right, and I fell out head first. Anyway, my parachute opened and all the strings were around me. I was turning like a top in the air. Thank God we were dropped from a good height and I had time to straighten out before hitting the ground. It was wonderful, quite exciting. From such a good height you are like a bird, you know.

"At home my mother told friends, 'I have three sons and the smallest [me] is the toughest.' She wrote me a letter after my first jump saying, 'I'm so happy for you Brigitte. You waited so long to do this.' From then on she called me 'mon petit aiglon' — little eagle."

Word spread quickly through the military and the press corps. Brigitte had earned their respect. Later, journalist Bernard Fall wrote that she was, "well known in Indochina since she was the only woman reporter in the whole command who held a military parachutist's diploma"[1] and added with respect that she "was one of the top combat reporters in Indochina."[2] Fine praise from a man who became known as the French authority on the region.

"I did become somewhat famous there, among the troops, Brigitte concedes. There were advantages, but sometimes discrimination. "When my stories were considered too technical the magazine would run them under a fictitious male name. My byline appeared otherwise." She seems to accept that that was the way things were in the 1950s in France.

"In general, I think it was easier for me because I was a woman." She would write, "It's true I excited a few dreams, simply because I was a woman."[3] She says, "I was alone, a single woman fifteen thousand miles from France with a lot of

French men. There were other French women there, in Saigon, in Hanoi, driving army ambulances. There was even one woman, Captain Valérie André, who was a doctor, piloting one of the few helicopters we had there. But there were none out in the field where I was. The men couldn't resist showing off in front of me. In the headquarters they could compete, give me extra information, just to have the pleasure of having dinner or lunch with me. It was strange. I heard colonels and generals telling me secret information just to impress me. Of course they knew they could trust me not to print the information. But I always got leads on stories." In the city she would be "dragged to dinners and nightclubs in full view of envious pals. White girls were a rare commodity. But that doesn't mean to say I found it entirely unpleasant."[4]

As for the soldiers in the field, she recalls that they acted with both consideration and friendship. She wrote that "to pretend that soldiers had the same relationship with me as with male war correspondents would be ridiculous and untrue. There was often silent attentiveness. Delicate gestures from boys who were often coarse."[5] She adds, "Out in the field, in the jungles, I was with the men. I lived with them. We slept out in the open or under canvas. People always asked me if that didn't cause problems — being the only woman among all those soldiers. It is ridiculous." She wrote, "Those who have never lived in the fraternity of war never seem to understand that I was seen as a female only when my companions had time to think about it. That didn't happen often. The rest of the time when we were moving through the jungle or rice field, in the humid night, or under the hot sun, when we were sleeping dazed with fatigue, I was — I could only be — a friend. Anyway, their thoughts were elsewhere."[6] She adds that there was safety in numbers. "The only time I had problems with men was when I was alone with one. Never in an operation. I slept in among the boys, so how could one of them try anything? I was safer than Queen Elizabeth in Buckingham Palace."

The climate was exhausting, and danger constant. Asked about fear she claims that "I didn't think too much about danger. As a war correspondent the whole idea is to go where there is something going on. You don't think about it." But she remained sensitive to the suggestion that her presence might be disruptive or harmful. "Out in the field it was life or death, and no one can ever say that I held up the troops or hindered their progress. I saw a lot of men fall down from exhaustion, but I was always on my feet. I had to be.

"Before one operation the colonel in charge of intelligence, who was a friend of mine, said, 'Brigitte, you can't go with this intelligence commando. They are going for three or four days inside the Viet Minh zone, and it would be too hard for a girl. You don't want to think that if something happened, or if they failed in their mission, it would be your fault, and I need the report.' I was quite moved by that, so I thought about it, hard, for about five minutes. I was so tough and well trained by then that I decided I could do it, and I said, 'I'm going.' He said, 'All right, then.'

"We had trouble as soon as we got in the Viet Minh zone. On the second day, after a quite difficult night, we were resting beside a brook, drinking and washing our faces, when suddenly a heavy firing burst out. We started to run through the bushes, so thick you couldn't see farther than a foot ahead. I heard the commanding officer, a young lieutenant, yelling, 'Where is the girl? Where is the girl?' But nobody tried to find out. They kept running and they were right to do that. I was running too. Everybody has to run out of an ambush.

"In the jungle the law of the muscle prevails," she wrote. "I weighed under ninety-four pounds, so when they could they helped me carry things or put up my tent. In that respect I was treated differently, and I appreciated that. But in life-or-death moments I was expected to take care of myself."

A woman reporting from the war zone has, one might assume, some unique insights to offer. Brigitte says: "Perhaps. I don't know. I think it is more of an individual thing, how you report. It depends on how new you are. The first time you see a comrade killed, the first corpse you see, is very moving. You are affected, and this may be reflected in the dispatches you send back. I was different in that I'd already seen thousands of corpses, on the roads through Czechoslovakia, in the concentration camp. I got accustomed to it. Not really immune, but . . . maybe in a certain way." There is a pause, then: "But you know, I can still remember the first corpse I saw in Indochina."

Sometimes it was difficult, she says, to remind herself that she was a reporter, not a participant. "I was out with men who were my friends. Sometimes I wondered if I was personally absolved from what I was seeing." On some missions they insisted she carry a gun. They gave her a pistol, saying, "This is not a stroll with a battalion we are offering you, with artillery protection and air support. This is a commando in an enemy zone." In her memoirs she described a three-day ambush she covered in 1952:

> The plain reached to within thirty kilometres of Saigon. Its tentacles of water and mud reached the border of Cambodia, an area always used as a safe den for pirates, rebels and partisans through history. Even American helicopters would not change that.
>
> We took off before dawn the second morning and by eight o'clock the ambush was ready. We were hidden in the elephant grass and bushes all along the track which bordered the canal. It was expected that one of our infantry battalions would be moving towards us, pushing ahead of them some Viet Minh.
>
> Even though the sun was high in the sky, we were forbidden to smoke, even to make the slightest movement that would stir the rushes. It was warm. The mosquitoes, though less ferocious than at night, were biting necks, arms, faces already drenched in sweat.
>
> We waited. I was hoping the ambush would work, it would make a good article. The day before we had moved for ten hours, and all this morning through water, water up to our calves, up to our shoulders over our heads — so far, for nothing. Not even the smallest engagement.
>
> But I also worried that it would work. It seemed rather disgusting. These 180 concealed men, machine guns ready, for a few guerrillas in flight. I hoped that a unit the size of ours would appear. Then I'd have my story, as the Americans say, but wouldn't be taking part in a massacre, just witnessing combat. I preferred a clear conscience to the absence of risk.
>
> By nine o'clock the heat was torrid. It was difficult to remain motionless and I wanted a cigarette badly.

With the chief commando's binoculars, I saw the first silhouettes leaping. The prey. We would only have to wait until they fell into our trap. The heat was getting worse. Flies and mosquitoes more insistent. Streams of perspiration were running down our khaki clothing, already stiff with mud.

The silence had become total. No limbs quivered, no hands swatted at mosquitoes. The sentries glued to the ground. Dressed in the large black cotton pajamas of the Viet Minh, a man climbed over the dike, cautiously, half bent. Watchful. Then stopped. I wanted to yell, look out imbecile, screw off. But to what good? In any case he was already in the net.

Two commandos jumped him. He protested, showing a used card of the 309th Regiment of the Viet Minh, but saying he had deserted the army two years ago, that he was only a poor, innocent hairdresser. The long knife in his belt only a traveller's knife. The grenade in his pocket found on the track. Maybe he was telling the truth. Was he a scout or really a civilian? It didn't matter, they wouldn't let him go. I expected them to tie him up. But after the questioning ended the lieutenant nodded his head, and the man was dragged away.

I thought surely I'd misunderstood this silent dialogue. It was not possible. The officer wouldn't let it happen.

Then, the sound of a gun.

"We never take prisoners," the lieutenant tried to console me. "One of us who falls in Viet hands would suffer the same fate. It's the law of the plain."

Maybe it was the delirious heat that made me suddenly feel like vomiting. I thought of the little black heap on the side of the track that the killed man would be, like the hundreds and hundreds of my companions in the past, on the roads of Czechoslovakia. Only this time I was chatting with the killers. By my presence alone I had taken part in the killing. What beastliness. . . .

I can still see the face of the liquidated prisoner.[7]

By February 1953, after two years of combat reporting in the field and frenzied social life in the cities, Brigitte began to feel the effects of the climate and pace. She took a three-week break and went back to France physically exhausted.

Two months later Geneviève de Galard, a Parisian nurse, arrived in Vietnam. Though at twenty-eight she was only two years younger than Brigitte, the two women could not have been more different. Geneviève had also grown up in the dark war years, but she had experienced France's fall and occupation in much less personal terms. She spent the war cloistered in a Catholic girls' school in the countryside near Toulouse, later nursing at a local clinic. But those years had left their mark on a person of sensitivity and deep religious conviction. She remembers, "I cried and cried when the war came, thinking how terrible war is, and how terrible it would be for France. I have vague memories, taking cover from falling bombs, feeling as though it would last forever. And finally, the joy of liberation in Paris." After the war ended, the slim, unconsciously elegant, blue-eyed brunette gave up plans for university and instead turned to nursing. She badly wanted to volunteer for Indochina as many of her friends were doing but did not feel that she could leave her aging and often ill mother. Finally, a friend told her that The Medical Service of the air army had a women's corps that supplied nurses for the medical evacuation planes in Vietnam. The job suited Geneviève because the nurses involved were sent to Indochina for six months and then rotated to posts in Africa. En route they could stop in Paris, and so she could get back to see her mother frequently.

Her mother, a viscountess, did not fully approve of Geneviève's plans and was frightened by the thought of her flying. But nursing in war was considered an honourable profession in France, even for one of the oldest and most respected military families whose ancestors had fought in the Crusades and served with Joan of Arc. Above all the de Galards knew about honour and duty to France, and this sense of duty placated the viscountess as it motivated Geneviève: "I guess it was a hereditary thing. I thought, 'I am young and free, and I should be helping the effort there.'" Despite the controversy the war had caused in recent years in government and the media, Geneviève never questioned her duty to France, or France's motives. She believed and continues to believe that "it was no longer a colonial war, but a fight for freedom — against Communism."

Once she arrived in Hanoi, there was little time for the shy but friendly woman to adjust to the completely foreign life there. For the next six months she flew daily evacuation flights to combat zones in air army Dakota planes equipped to hold thirty-two wounded soldiers. Her first missions were to combat operations in the north of the country for Operation Castor and in the region of Tut Hoa in the south for Operation Atlante. It was her first exposure to combat, "but I wasn't thrown into the battle when I first arrived. We usually landed a few kilometres from where the fighting was, and there was no real personal danger. But some of my friends had already been wounded. I guess you always think it won't happen to you. I was more terrified for my girlfriends, when I knew they were going to a dangerous zone." As for herself, "I must say, I was very eager to accomplish my missions."

Brigitte returned to Hanoi, "happy to be back and rested" but fully aware of how bad the situation was becoming. The French had lost most of northern Vietnam and a large portion of Laos. Cambodia, though now declared independent, had become a supply route for the Viet Minh. She spoke with her friends and contacts and scouted out the next big story of the war that had by now cost seventy-four thousand French lives. "I felt by then that the war was lost. We had won the large battles when we could confront the Viet Minh face to face, but as I wrote in my notes — and this was in 1953 — 'From victory to victory we are going to be beaten.' We couldn't win that kind of war."

In France the war had become increasingly upopular and politically divisive. There had been fifteen changes of government in the past 8½ years. Returning veterans were experiencing the unpopularity in very personal terms, often finding that coming home was as painful as the experiences they had left behind. As the cost in lives and francs grew, so did popular impatience and questioning of the war's worth. The government wanted an end to the war but also some clout at the bargaining table. Before negotiating a settlement, it wanted a military edge.

In May 1953 General Henri Navarre was appointed new commanding officer of the French Forces. Though he had an impressive military record in World War II as French Chief of Staff with the Allied Command in Europe, he had no experience in Indochina or with guerrilla war. His assessment of the military situation was grim, and as he later said, "France in any case was tired of the war." He therefore decided to lure the Viet Minh into a battle far from their bases that they could not easily supply. There, supposedly, French air- and fire-power superiority could prevail.

The setting he chose was historically significant to the Vietnamese. Dien Bien Phu, also known as the "arena of the gods," is an opium-growing valley in the northwest corner of the country, only six miles from Laos. It had been the scene of epic battles in centuries past, fought over for its strategic value as a converging point for most trade caravans through Asia. A village of only five hundred families, it was a burial ground for many thousands more. The first French troops were dropped into the area on 20 November 1953. Even that first day was bloody; unexpected Viet Minh companies shot at the paratroopers as they fell from the sky, but by nightfall the French had secured the valley. On 22 November Brigitte, carrying extra fatigues and cigarettes and a newspaper to read on the two-hour flight, joined the last paratroop battalions that were to make up the initial garrison in the valley. She was ready for the big story — the final stand at Dien Bien Phu.

"On board there were about twenty-four people plus the dispatcher and the crew. We were all a little nervous, I think, but I took out my nail file and did my nails while reading the news. A sergeant near me was in a state of complete panic and ended up pulling open his parachute. The dispatcher saw him, and it was really awful. He started to yell at him, 'Look, there is a girl in front of you, smiling, doing her nails — and you, poor bastard!' and he slapped him to get him out of his panic. I was very embarrassed, but I knew it was important to hit him, to knock him out of it, because if the young sergeant had refused to jump he would have faced a court martial for certain.

"I had a bad landing. I lost my under-helmet and stumbled on a corpse. I turned around and there was this sergeant saying, 'Look you, here. I jumped.'"

Brigitte began covering the build-up on the plain, which measured approximately seven miles around. Sappers tore down the village houses and built bunkers and a command post. They erected seven outer bastions, which would each house an artillery and infantry regiment. In the centre of the camp an airstrip was laid. The furthest outpost, Isabelle, went up about five kilometres from the centre of the base and was surrounded by barbed wire. The entire camp was underground, with covered and uncovered trenches leading to the underground rooms. French forces, which would reach a total of fifteen thousand men, would be supplied entirely by air.

Geneviève began her missions into Dien Bien Phu a few weeks later, in early December. "At that time it was quiet there, so we'd land and spend a few hours, sometimes having lunch with the officers. I had time to see the camp and visit the surgeons and male nurses. The trenches of the hospital were covered with wood and sandbags, and the surgery room's mud walls were covered with white material. Altogether there was room for about forty wounded.

"We used to wait until evening to take off, waiting to the last minute for any more casualties, and then load them on the plane and go back to Hanoi." Most of the soldiers evacuated were suffering from illnesses such as typhus. Some were wounded in field operations the French were still launching from the garrison at Dien Bien Phu, deep into hostile territory, before the camp itself came under attack.

Brigitte accompanied the men on some field missions from Dien Bien Phu, preferring them to the "never-ending weeks in the holes" of the camp. She was parachuted alone into Muong Khoua, a hard-won base in the mountainous border with Laos. "I was dropped alone, in the mountains, at dusk. I landed on a very slopey area, covered with trees and mined. It was very difficult to get me out of the trap. My face was covered in sweat even though it was not hot. I washed it, trying to compose myself before meeting the officers in their HQ. I remembered that I had some lipstick in my field bag and decided to put it on. I must say that everyone was very shocked with my jump. A few days later, the commanding officer of the battalion I jumped to meet told me, 'Brigitte, you astonished us. Not the jump — we know about you — but when you appeared in the CO tent with lipstick on!' After that I always wore it. I knew it made them feel good. In the air force headquarters in Dien Bien Phu they pinned up a cigarette butt with lipstick on it — mine, of course."

From Muong Khoua, the forces made a twenty-five-mile trek on foot to another base. Finding it had been overrun, they started back to Dien Bien Phu. French intelligence had determined that there were four Viet Minh divisions in the area, far more strength than even the most pessimistic had envisioned. In his book about Dien Bien Phu, *Hell in a Very Small Place,* journalist Bernard Fall recounted: "The return to Dien Bien Phu was even worse than the approach march. Unseasonal rains had fallen and the already swift mountain rivers were even harder to cross. Christmas was spent in wet uniforms in the middle of the jungle. . . . Finally on the morning of December 26, a cry went up among the lead elements of the column: 'Dien Bien Phu!'"[8] Fall goes on to recount that as the French crossed the last hill an isolated rifle shot was heard and a paratrooper killed. The valley was surrounded by Viet Minh. He adds, "As the exhausted paratroopers stumbled into camp, Brigitte Friang noticed a 'Christmas tree' in front of one of the dugouts — a tree that might have come straight out of Goya's "Disasters of War." The trunk was a pole used to hang barbed wire; it was crowned with an enemy helmet and carrying for ornaments a well-worn GI boot, empty bottles, and smashed cans of army rations."[9]

Movements in and out of the camp were becoming impossible. The guerrillas were digging into the wooded hills that surrounded the base. Against all odds they had brought in 105-calibre heavy guns, using a convoy of six hundred Chinese trucks, and then pushing the "steel elephants" up the last fifty kilometres of thick jungle on foot. A human convoy of twenty thousand men on foot or bicycle transported food and supplies. General Navarre, who had bet on the Viet Minh's inability to supply itself, now worried about his own men's complete dependence on air support.

Brigitte spent three months in the dreary, muddy camp. "We slept in dug-outs, and in the morning when I wanted to wash the boys pinned up a blanket for me and left me alone for a few minutes with a bucket of water. After I finished I would just call out, 'You can come in now.' They tried to give me privacy. Sometimes the lack of it made life difficult. Being a woman meant it wasn't always easy, because of feminine troubles, but I managed."

By late January "nerves were frazzled." The troops were ready for the battle and an end to the siege-like atmosphere.

On 21 January Brigitte celebrated her thirtieth birthday in command headquarters, where they drank a lot of champagne and talked about the impending battle. "I told the officers that my editor was starting to become impatient and that he wanted me to leave. He kept sending me telegrams saying, 'Brigitte, come on, there is no story there.'

"Finally, a few weeks later, I received a telegram from him demanding that I leave. I went to say goodbye to General de Castries. I said, 'I'm sorry my dear, but I have to leave,' and he said, 'No, Brigitte, don't go. We want you here during the battle.' I remained convinced that he wanted a civilian to witness the attack. But of course I had no choice. I decided to go to the Delta to cover the big battles that were happening there and to return to Dien Bien Phu on the day of the offensive."

The attack came on 13 March 1954, only one day earlier than Brigitte had predicted. Viet Minh artillery opened fire on "Beatrice," one of the seven outer bastions that formed a chain of strongholds around the camp. Viet Minh suicide squads cut through the barbed wire and blew themselves up with their grenade attacks. By midnight Beatrice had lost five hundred of her seven hundred men, and two other outposts had fallen.

Friang's colleagues parachuted in the next day. She had planned to be with them. One was killed before he hit the ground; the other lost his leg. "I keep telling myself that if I'd gone with them there is a good chance I would have met the same fate, and yet I still can't forgive myself for not going. I had phoned to confirm my departure on board the carrier only to be told I could not go. I was told that General Navarre was arriving momentarily, and that they couldn't allow me to go without his approval.

"Navarre gave me a flat refusal. He said, 'No women at Dien Bien Phu. They are evacuating Colonel de Castries's secretary.' I begged him to phone de Castries, who I knew wanted me there. They refused, too sure of his answer. Out of rage, disgust and confusion, telling myself my companions would think I'd abandoned them the moment things started going badly, I left

Tonkin and, a few days later, Vietnam, where I was no longer sleeping because of my anguish."[10]

Geneviève and the other flight nurses continued flying into the camp to evacuate the wounded, but now the flights had to be made at night and the planes landed only long enough to pick up the wounded. "When you landed it seemed as though you were landing in hell," she remembers. "There in the darkness lay the men waiting to be evacuated, staring at you with sunken eyes from pale, exhausted, unshaven faces.

"One night we went in, landed, but the ambulance wasn't there. The artillery started and the pilot decided he couldn't risk waiting and started down the runway. Just then I saw the ambulance coming toward us, and I rushed to the pilot and said, 'Stop, stop, the ambulance is here!' But the artillery was going off and he said, 'It's impossible,' and took off anyway.

"The next night was the twenty-eighth of March. We took off from Hanoi at four-thirty in the morning and arrived at Dien Bien Phu at around five forty-five. The visibility was very bad and the pilot couldn't see much of the runway because there were only two small lights at each end. He tried to land three times and on the last attempt hit the runway off-centre and crashed into some barbed wire and spikes. We didn't realize that the spikes had perforated the oil tanks so as soon as we landed we began loading the wounded onto the plane. Because of bad weather, these poor men had been waiting to leave for three nights. When everyone was in the plane the mechanic came and told us the bad news. It was impossible to take off in that condition, and we would have to wait until it was fixed. We had to unload the wounded, which was very depressing. They thought it was the end of that hell for them, and it wasn't.

"I went to see the doctor and asked him to promise me that these same men would be the next to get out when a plane finally arrived. He said he couldn't promise me that because he had to give priority to the most severely wounded.

"As the sun rose and the fog lifted, the Viet Minh opened fire on the Dakota. From where I stood at the command post, I could see the plane hit by a shell and burst into flames. It was very emotional.

"I went back to the hospital and placed myself under Chief Surgeon Dr Paul Grauwin's orders. I didn't expect to be there long. More planes were expected, and there were strict orders not to have any women in Dien Bien Phu. That night we got in a jeep to go to the airstrip to wait for the planes. I had packed my few things and was carrying a lot of letters from the soldiers to mail in Hanoi. The atmosphere had changed. One of our ammunition dumps had been hit and a huge fire was lighting up the sky. It was bad weather. The ground was wet and our jeep got stuck in the mud so we got out and walked. No planes arrived that night."

The next day, 30 March, at five in the evening the Viet Minh began a massive shelling and attack of the camp. In another of war's many twists of fate Brigitte was not there for the story of a career, but Geneviève was stranded. Brigitte was later told by Navarre's second in command, General Cogny, that she was stupid to have left the country. He said that as soon as Navarre left Hanoi he would have allowed her to parachute in with reinforcements, but, Brigitte asked, "Who would have believed that with the critical nature of the war in the Tonkin and Dien Bien Phu, General Navarre would go quietly back to his palace in Saigon, at the other end of the peninsula?"[11]

Within hours the Viet Minh had captured three French outposts. The fighting was hand to hand in the trenches, Viet Minh soldiers falling over machine-guns in suicide attacks. Geneviève waited with two surgeons and the male nurses in the hospital trenches. "During the night it was impossible to bring the wounded to us, but by dawn, about five in the morning, hundreds of wounded men began to arrive. Our hospital, built to hold forty men, was submerged with stretchers."

The following night the French launched a counter-attack. Lighting up the camp with flares, they heavily shelled their former positions. They recaptured only one. Continual explosions rocked the dirt-covered dug-outs of the field hospital. A shell took out the electrical lines, and the tunnels and surgery rooms were plunged into darkness. Geneviève worked with a flashlight, climbing over the stretchers of wounded, which soon were lined up along the tunnel leading to the surgery room. She could barely find room to kneel down in the mud beside them to administer morphine or some other anaesthetic. "By five o'clock in the morning the wounded could no longer fit inside the catacombs of our hospital. Outposts Dominique I and II had fallen and we were overwhelmed with dying men. Some were left lying outside the trenches, unprotected from the shelling, and many were wounded again or killed as they lay there. It was morally terrible — I couldn't even look at them. Dr Grauwin was stumbling over the bodies, trying to decide which one to operate on next.

"Suddenly, there was silence. I learned to my amazement that three days and nights had passed." Four thousand French soldiers had been wounded, and nearly two thousand were dead. "I walked through this scene, leaned against the wall and closed my eyes. At that moment I wished that I could just go to sleep and never wake up again."

Geneviève found a damp, dirty cot, unfolded it among the wounded, lay down and fell asleep.

The Viet Minh, too, had suffered extreme losses, and now for a time there was quiet. Soldiers began clearing out their dugouts to make room for the wounded. They dug niches in the mud to form beds when the cots ran out. The airstrip was no longer in French hands, but Geneviève says: "I felt I could not have left anyway. Not after those three days. I wouldn't have wanted to go."

She sent a telegram to her mother saying, "The boys have invited me to stay for the battle." She laughed when her mother sent back the reply: "I am relieved. At least there you will not run the risk of being killed in a plane crash."

As the doctors continued to perform up to twenty-five operations a day, Geneviève organized the post-operative cases who could no longer be evacuated. She began her work each morning at five o'clock when casualties arrived from the outposts. At seven she distributed breakfast. Food was a problem for many of the wounded, so she approached the cooks and asked them to prepare soup or something easier to digest.

Her work involved changing dressings on patients with abdominal injuries and colostomies, which meant replacing gauze soiled with feces. She changed the dressings on painful gaping wounds. Dr Grauwin later wrote that she was the only one who could treat some of the patients without their crying out.[12] Geneviève says, "I think men and women have different ways of taking care of sick people. Women are more sensitive, perhaps. Even if I was sometimes obliged to be a bit rough and they cried out, 'You hurt me!' I did my best to be kind to them."

The hardest adjustment for this quiet, aristocratic woman who says she "lacked self-confidence" was the complete lack of privacy. "I was filthy," she sighs. "The men weren't allowed to use the distilled water for washing, but they permitted me a basin. The hardest thing was going to the bathroom. I had to climb out of the trench into the open, behind the sandbags and barbed wire. I was very frightened that I would be wounded, and I really didn't want to be wounded in that position!" She laughs, but the lack of hygiene bothered her. The men often urinated into the empty flasks that were around and then flung the contents outside.

Geneviève still marvels at the men's concern for her and remains touched by the small gestures they managed to make. "One day, about three weeks after the main attack, Colonel Pierre Langlais, Commander of the Paratroopers, asked the doctor where I was sleeping. When he found out it was among the wounded he secretly ordered his men to prepare a small shelter for me. They covered the mud walls with silk from a parachute and found an armchair and a real cot from the command post. He came one day and said, 'Come Geneviève, I have something to show you.' And then he presented me with my new room."

It is said that the male nurses, who also slept among the wounded, were upset at this gesture because they hadn't been able to do it for her. Dr Grauwin later recalled that they felt she belonged to them and sometimes became jealous when they were "shown up" by the officers. They offered her what they could, and soon she had a few more pairs of trousers, some shirts and a paratrooper's camouflage suit. She had arrived with only the khaki trousers and white blouse she wore — and one tube of lipstick, which she used on "special occasions, to feel a little like a woman."

Having a room of her own made her life easier. "I was so touched. I was so happy that now I could sometimes be alone. It had been terrible not having any privacy. This meant so much to me."

When there was time, in the evenings, the hospital staff would sit and talk with the wounded. "One of the male nurses was a very funny boy from the north of France, and he would amuse the wounded with hilarious stories. He always made us laugh. Sometimes we sang songs."

Each of the four remaining outposts had its own surgeon and male nurses. Geneviève insisted she be allowed to visit them. "I wanted to take them some oranges, condensed milk and cigarettes that had been dropped in. I took my helmet and made my way around the uncovered trenches. The men were all amazed to see me — a woman — there. They didn't know I was in the camp. I really enjoyed these visits, but I think it scared Dr Grauwin. He was always saying, 'Geneviève, don't go so far, there are shells going off everywhere. I don't need another patient to care for!'"

In his memoirs, *Doctor at Dien Bien Phu,* Grauwin devoted a chapter to Geneviève: "She had almost made a complete circuit of the main camp. Shells had been falling all the time, and she had to take a very complicated route, going out of the trenches across paths in the open under fire, jumping over shell holes, and forcing a way through barbed wire. . . . It was nearly two hours before she came back, flushed and out of breath, covered with mud, and, to judge from her face, in the best of spirits, on top of the world."[13] The doctor had served in Indochina for many years and had witnessed the courage and sometimes the casualties among the French army nurses. He wrote: "It was then I realized that she was entitled to a place in the great procession of extraordinary young women which went on without a break for more than eight years in Indochina."[14]

Geneviève remembers happy moments, particularly the dinners she spent with the officers. "It helped my spirits a lot to see men who were healthy. I would wear my paratrooper suit and put on my lipstick. I think they enjoyed it too, to have a woman to talk to. I might have made it seem a little less awful for them. Then I would come back to the hospital and pass on the latest news to everyone there."

News was not good. The Viet Minh were digging towards the centre of the camp, constructing a network over a hundred kilometres in total length. The sounds of their shovels could be heard inside the French command post, and French bombers could not help. General de Castries and his commanding officers kept hoping that there would be a massive bombing campaign launched from Hanoi, with help from the Americans. In fact a plan had been drawn up for a two-hundred-U.S.-bomber offensive in support of the beleaguered garrison, but Congress voted against any further expenditure. Senator John F. Kennedy, in a speech to the House, declared that the French had no chance of winning a military victory against "an enemy of the people, who were supported by the people." Senator Lyndon Johnson agreed.

While the Viet Minh dug closer to the camp, French sappers dug new trenches to house the wounded. They finally stopped when they reached the old burial grounds. Underground the nightmare continued. "The heat and humidity were becoming terrible as monsoon season approached," says Geneviève. "That made it much more uncomfortable for the wounded. Sweat trickled down inside their plaster casts, and so did the mud, making the wounds difficult to heal and very painful. The trenches were so filled with mud that it was almost impossible to keep things clean, and by now there was very little filtered water left. We did what we could for them, but we were losing a lot of supplies in the parachute drops that were falling over enemy territory. We were short of plasma."

One day she heard screams and, rushing to help, saw "large white maggots crawling in the patients' bandages and wounds. Dr Grauwin tried to comfort the men, saying, 'Don't worry, they are good. They prevent gangrene.'"

"I kept my spirits up by telling myself that I didn't have it bad compared to the wounded men. The worst times were when the men were dying. I could stand the maggots, the spiders, the mud — everything. But when they were dying I sometimes thought I would go out of my mind. I remember a young officer

Natalya Kravtsova

Nadya Popova (standing) and other pilots of the 46th air regiment rest in camp

Novosti

Soviet pilots report after combat mission

Haganah fighters, Negev settlement

Haganah soldiers in training

French Army photo

Geneviève de Galard, Dien Bien Phu

Courtesy Brigitte Friang

Brigitte Friang in Vietnam, 1968

Courtesy Mary Beth Crowley

Mary Beth Crowley in front of her Quonset hut

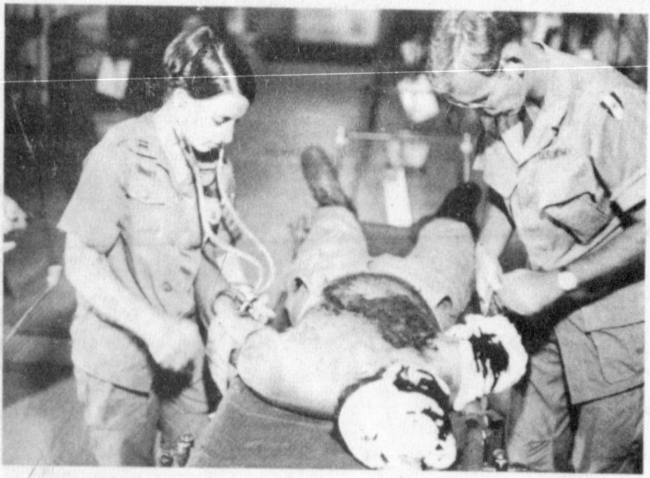

U.S. Army Archives

Triage

Minneapolis Nurses Project

Memorial statue of American army nurse, St Paul, Minnesota

Brigadier Tony Wilson

Linda Kitson

AMES Photo

Salvadoran guerrilla

Novosti

Soviet veteran of World War II, May 1984

who had been paralysed, and the paralysis was increasing. He was finding it difficult to swallow, to breathe, and we knew he was going to die. I was sitting with him, trying to help him swallow a little soup, and he kept begging me, 'Promise me I will not die.' It broke my heart. Another, wounded and blind, asked me to read out loud the letters from his wife. I had to pass over certain parts because he was crying."

Her religious faith helped, she says. "It was so important to me during the battle. I always attended the masses given by the army chaplains. When I was distraught I turned there for strength. My faith did not waver during those terrible times. God gives freedom to men and cannot be held responsible for men's actions."

Whatever her personal fears or despairs, Geneviève knew that she must be cheerful for her patients. She was not the only woman at Dien Bien Phu. There were prostitutes from two local brothels, and some families of Meo tribesmen among the twelve thousand soldiers. But she never saw another woman. Bernard Fall wrote that "she was a modest girl . . . with a ready smile. In a world gone mad, such . . . remnants of normality did a great deal for the morale of the garrison."[15]

"The men thought of me more as a mother than a woman, I think," she comments. "They used to call for me when they had some improvement or if they needed some extra care just to have me comfort them. As for the soldiers, I think I raised their morale too. I was very frank, and clear that I was there to help — not to flirt. We all lived together and there was never a time when one of them laid a hand on me. Moreover, none of the others would have allowed it."

Up until the end reinforcements were sent in to Dien Bien Phu. Six hundred and eighty men with no parachuting experience volunteered to be dropped into the siege. "It was very brave of them," says Geneviève. "Many hurt themselves landing. One night we watched from the tunnel and saw all these white chutes in the sky. It gave us hope to see fresh troops from Hanoi and to know that we weren't forgotten.

"I remember one young soldier landed near me, and rising to his feet he said, 'Well, a woman here! I guess this isn't hell after all.' It always amazed me to see the spirit of the men, especially the wounded. One boy who had lost both arms and a leg was one of my special patients. When I could I used to take him outside to get fresh air. He would lean on me with his stump, jumping along the tunnel on one leg. He told me, 'When we are free I will take you dancing.'

"Something very special happens among men who are thrown together in very dangerous situations. It's a deep kind of friendship which is difficult to describe. I saw men volunteer to go fetch the wounded in the most dangerous situations. The surface of the camp was covered in shell holes, the roofs of the trenches were caving in, no more jeeps were functioning anywhere, and they would go back out, often getting wounded themselves. It is remarkable for a woman to experience and share in this kind of thing, and I found it very moving. I was living through such stirring moments, as a witness to men's courage. At the same time I was gaining more confidence in myself. The pain was maturing me."

Typically, Geneviève did not think about her own reserve of courage or will-power. But it had not gone unnoticed. On the night of 29 April Colonel Pierre Langlais invited her to headquarters. "When I arrived everybody was there, including General de Castries. I never saw the General normally, only when he came to decorate the wounded. He stood up and said, 'Geneviève, I have been waiting for you. I have something for you.' He handed me a simple white envelope. I opened it and discovered the Legion of Honour and the Croix de Guerre. Then he took them from my hands and pinned them on my shirt. Imagine my surprise and emotion. They couldn't get decorations from France, so they had given me their own. The Legion of Honour had belonged to one of the captains, and the Croix de Guerre was Colonel Langlais's. The Colonel told me, 'Many men would not have been able to do what you have done.'"

Though she tried to be hopeful, Geneviève could not help believing the battle was lost. She says, "It was impossible to win because the Viet Minh had such strong artillery. The only hope was to hold our position until the rainy season came, or until the Geneva Convention began." The peace talks at Geneva were scheduled for 8 May, but by 30 April the Viet Minh had dug their way almost into the camp.

In the catacombs in the central camp the Cameroon Legionaires celebrated their anniversary that night. Geneviève was the guest of honour and was made an "honourary Legionnaire, first class." She says, 'I felt so proud to be considered a soldier — one of them." When she returned to the hospital, Dr Grauwin suggested she show her decorations to the wounded. She refused, saying, "I have been given all this, but what do they have?" The doctor opened the last ten bottles of his champagne and the male nurses congratulated her. Finally they convinced her to tell the wounded. Distributing champagne, wearing the green and black shield of the Legion, she said, "I am like you now." At which, the doctor wrote, "There was an endless chorus of approval and congratulations."[16]

On 1 May the final attack began, and the fighting did not stop. "We were again flooded with wounded. Some of the outposts were overrun, and they tried to get their medical personnel and wounded to us. Some did make it — those who could still walk, pale as death, covered in mud." Others returned who had only recently been allowed to leave the hospital, men without an eye, without an arm or a leg, who had gone back to fight. Now they returned, with mud thick under their plaster, some with new wounds.

On 3 May General de Castries appeared in the hospital to decorate the wounded men. Unfortunately, he told them, the medals being sent from Hanoi had been parachuted over Viet Minh territory.

"By the night of the sixth, we kept hearing enormous explosions, as one by one our posts were falling. I'll never forget that night. The fighting was incredible; it seemed endless and beyond human endurance. On the morning of the seventh I went to the hospital at five o'clock to meet the incoming wounded. I stayed there all day. Around four in the afternoon I went to the command post to get some information. They were discussing what to do — whether they should turn out all the soldiers for a fight to the death, for honour. When Colonel Langlais saw me he said, 'Stay here, Geneviève. You will be safer with us.' I lay for a few minutes under the table, just to have a few moments rest. I shut my eyes, shutting out what was happening. As I lay there I heard a voice over the radio. It was one of the commanders of the outposts saying, 'The Viets are here, only ten metres away. I send a kiss to my family. Goodbye to everyone. This is the end for me.' That was the most agonizing moment."

The final orders came from Hanoi. Cease-fire was ordered for 5 P.M., exactly fifty-five days to the minute after the battle had begun. Soldiers began to destroy the ammunition and burn documents in headquarters.

"I returned to the hospital and broke the news to the wounded. They didn't say much. I began to give out the last cigarettes. Some men were quietly crying. Many were worried about me. Others tried to reassure me, saying the Viet Minh would not harm me.

"All the staff met in the tunnel at fifteen minutes past five. We heard the sound of shoes above our head. A Viet Minh appeared at the entrance and said, 'Get out.'"

Outside the ground still smouldered, ammo dumps burned and the earth was pock-marked by shells. Trenches had collapsed and filled with mud. Jeeps and trucks lay overturned. Dead bodies, half sunk in the mud, were swarming with flies. Dr Grauwin warned Geneviève to stay close to him as they were marched through the camp. The sky was a brilliant blue; the sunlight hurt their eyes. The air was acrid with the stench of smoke and death. An orderly stared in shock at the doctor in the sunlight and said, "Doctor, your hair is completely white."[17]

A Viet Minh commander approached them and told Geneviève and one male nurse to return to the hospital. "I took a long look at the camp, at all of our defeated soldiers. I was prostrate. It had been such a courageous battle, our men were surrounded and fought on without much hope. Hanoi knew the situation was desperate, and so they didn't send all their troops. But people like Colonel Langlais had kept hope until the last."

The two nurses spent the next three days caring for the severely wounded men who could not be moved. "It was very depressing. The Viet Minh had taken all our medicine and dressings for their own wounded. I had managed to hide some large dressings for stomach wounds, but that was all. We worked in semi-darkness, without distilled water. I think that is when I got amoebas."

"Finally they came and said we could move outdoors. They put up tents for us. We settled by the river in a special tent where we could treat the wounded with their equipment, and under their eyes. I think they watched me to make sure I was a nurse and not a spy." Geneviève described her captors as "arrogant" but correct in their behaviour towards her.

Living outdoors again and being able to wash herself in the river gave her "the impression of holidays. It was wonderful to be outside." But the rest of the world waited and worried about the fate of the prisoners, and especially of Geneviève. The press headlined stories about the "Angel of the Siege." She declares that the men never called her angel, "Only mademoiselle, or when they knew me, Geneviève. I think that name was invented by American journalists," she laughs.

The most frightening experience during captivity was "when the Viet Minh commander told me I had to sleep in their girls' tent. For the first time I became petrified. I don't know why. Maybe because of the pressure of propaganda. Already they had been telling us that France was wrong, and responsible for the conflict. They made us watch films. But the male nurses spoke up and said, 'She was with us in the bad times, we want her with us now.' Luckily they agreed."

They did, however, ask the internationally famous "Angel" to write President Ho Chi Minh birthday greetings. She says, "I was in a dilemma. If I didn't do it I was afraid they might delay the release of the wounded prisoners. So I managed what I thought was a clever compromise. I sent him wishes for peace on the occasion of his birthday."

Negotiations began in Geneva only one day after the fall of Dien Bien Phu. The siege had taken three thousand French lives, and an equal number of men were permanently disabled. Four thousand more were to die en route to the POW camp or during captivity. The Viet Minh had lost eight thousand men. They gathered the bodies of French and Vietnamese soldiers and buried them together in a mass grave in the valley. Above it they raised a bamboo and silk cenotaph.

In the settlements that followed, the French agreed to withdraw from Vietnam. The country was to be divided at the 17th parallel, with Ho Chi Minh governing in the north and a democratic government installed in Saigon. The agreement provided that in two years a national election be held and the country reunited under an elected leadership. Those elections would never be held, and the war in which Brigitte and Geneviève were witness to so much suffering would merely change hands and continue for another two decades.

The Viet Minh were embarrassed by the international press coverage of Geneviève's captivity. They decided to free her before the others, but she demanded to stay with the wounded. "At first they accepted my request," she says, "but later they came back telling me, 'We must force you to leave because the international press is claiming that we are holding you by force.' I was escorted to the plane by Dr Grauwin and my faithful assistant, Julot. The last words Julot said to me were, 'What are we going to do now that we have lost your blue eyes?'"

On 24 May, after three weeks in captivity, Geneviève boarded a Beaver plane for the French base at Luang Prabang in Laos. Her legs were swollen with beriberi and she had lost eighteen pounds. She felt "aged."

A French radio chief corporal stationed at Luang Prabang was there at her arrival and wrote to her about it thirty years later:

> I was there when you got off the little "Beaver." You were wearing, I believe, a paratrooper uniform. You appeared tired, a little lost from that incredible thing that just happened, happy just the same to return to French lines. You were very touched, I think, to inspect that troop of Legionnaires of the IREC stationed in Luang-Prabang, which formed an honour guard for your arrival at the time of your descent from the plane. It was very moving to see how everybody was looking at you and was honoured by the arrival of the little French woman.
>
> I think you were given a place in one of the U.S. tents of the two surgical wards set up in the border of the airstrip where all the injured came after they were released by the Viet Minh.
>
> Do you remember a radio chief corporal who asked if you

would like to send a telegram to your family in France? Well! That corporal-chief was me.

I remember that you drew up those telegrams sitting on the corner of the camp bed with my leather saddle bag on your knees serving as a table. You and I were surrounded by French and foreign photographers trying to capture the events on film. . . . Up until the time of your departure for Hanoi by Dakota, you were the focus of attention of everyone present.[18]

Geneviève's letter to President Ho Chi Minh had hit the press in an edited and misconstrued form. The woman who wanted only to "do my duty to France" had become a propaganda ploy, and that caused her pain and embarrassment. She was not then, and still is not, "a political person," and yet was used for political purposes by both sides. En route to Hanoi she was warned to "watch my words." When they landed at Gia Lam, "I got photographers' flashes in my face, and I remember I said, 'The hardest part has not yet started.'"

A *New York Times* journalist who waited at the airport described a chaotic scene as Geneviève descended into the crowds, and a Moroccan guard detachment took up position and "drove back the correspondents with fixed bayonets. 'How goes it?' someone yelled in French. 'Very well,' she replied, 'but I don't want to talk right now.'"[19]

"The authorities wanted to send me as quickly as possible back to France to escape the media pressure, and not to take any risks regarding the prisoners. I asked to see the wounded in the hospitals before I left. I spent two days in Hanoi visiting the men, and it was then I met the officer who was to become my husband, Captain de Heaulme, who was in charge of interrogating the released wounded soldiers."

On 1 June she arrived home in Paris and was met at the foot of the plane by her mother. She describes this period as "dream-like," but she found it difficult to fall asleep at night. She was suddenly a celebrity but was coping with disorientation and ill health. The United States Congress invited her on a three-week tour of the country in July. At first she refused, but later she accepted, after "French authorities explained to me the importance of going as an ambassadress for France." In New York she was greeted as a heroine and driven through the streets where 250,000 people came out to throw her a ticker-tape parade. In Washington President Eisenhower presented her with the Medal of Liberty.

Brigitte Friang had also become a celebrity in the United States. She gave an interview to a journalist in the press camp in Indochina just before leaving. "He was a friend, and I asked him not to print it until after the battle of Dien Bien Phu," she says. "He did anyway. When I landed in Los Angeles, I thought I was the president of the United States. There must have been forty or fifty journalists, photographers and newsreel people waiting for me as I walked out of the plane, flashing their cameras at me. Then all the journalists started to laugh. They were waiting for a very tall, strong woman — an Amazon — and out came this tiny Frenchwoman in a smart beige suit. They really laughed!"

An article in *Newsweek* described her as "a veteran of twenty-six months of battlezone reporting [who] bore delightedly little resemblance to a war correspondent, at only 5 foot 2 inches, and weighing a pertly feminine ninety-four pounds."[20]

Astonished by the impact she was making, Brigitte marvels that "In a few days, through television and newspapers, I was known from coast to coast, or as the Americans say, from the Atlantic to the Pacific. I needed police protection!

"And the people in the States told me, 'Just wait till you get back to France. You'll be a hero, it will be a triumph.' I said, 'Well, maybe my parents will meet me at the airport, but that is all. In France we don't make a big deal of these things.'"

She was approached by American publishers who suggested she write some steamy memoirs. "They said, 'If you don't have any, make them up,' she wrote. "A French woman alone in the exciting promiscuity of war and the treacherous jungle populated with little yellow villains with slanted eyes. Imagine. If I had agreed, I bet that within a month I would have appeared in the comic strips. As a blonde, taller, bustier, and with more hips, in a close-fitting shirt open to the navel and tight pants. The ideal bush look. A gun in one hand and a typewriter in the other. . . . I was a great disappointment for not welcoming these propositions that had the ring of a cash register. I was not judged as being honest, only as an idiot.''[21]

After she returned from the United States Geneviève spent a month writing to soldiers' families for twelve to fourteen hours a day. ''Psychologically it was a very difficult period. Sometimes I felt as though I were losing my memory. I was quite ill. The best solution was to return to my job. I went back to Indochina to help in the evacuation of soldiers and French families. It stopped me thinking about Dien Bien Phu too much. It would have been much more difficult for me if I had left the air army. A lot of the boys never recovered from Dien Bien Phu. They could never lead a normal life in France. It was such a difficult readjustment, some couldn't even find jobs. One of the male nurses died recently, and his family told me, 'He just let himself die.' It was very sad. For a long time I couldn't talk about it except to those who were there. I think the others feel the same way. Today all of us who were there remain very close. When we see each other we always embrace. I only realized later what it meant to them, to have a woman there. I was the human element they were missing.''

Brigitte: ''It is hard for people who have never experienced combat to understand the deep fraternity among soldiers that is normally reserved only for males. It is very deep and lasts forever. And I don't think soldiers are really a different breed or cut of person. I do think that in uniform, in battle, they change. The uniform changes people, and so they act differently, they follow orders because of concepts like 'duty' and 'honour' — to the death if they have to.''

Brigitte returned to France and eventually returned to her work with André Malraux. She married briefly, but now says, ''I don't think women like me are meant for marriage. I think that takes far more courage than I have, to become a mother and raise a family.'' By the mid-sixties she was back in Vietnam, this time with a television crew, reporting from the fire-zones, sending home agonizing images of the drawn-out war. Why did she keep returning to war? ''I would miss it. The experiences, the deep friendships, even the risk. Curiously, war taught me absolute respect for life.''

Geneviève married Captain de Heaulme and spent the next years raising three children and travelling to various army posts. It is only recently, as a grandmother, that she has given her first interviews. Being dubbed ''angel'' and ''heroine'' has put pressure on her. ''Because of Dien Bien Phu I am committed for life. I have to be faithful to the image people have of me, and this has been very restraining. I can never show myself to be an ordinary person, which is what I am. They have held onto the heroine image, and that forces me to try not to disappoint them.'' She pauses, then adds, ''People tell me I was courageous, but I think there are many kinds of courage. If I was brave, it was quite unconsciously.''

TWILIGHT ZONE
Vietnam, 1965-72

What the nurses saw in those operating rooms was an unrelenting procession of the bits and pieces of people arriving from the battlefields. What they felt, working twelve-hour shifts in a kind of "twilight zone" removed from the war yet dealing with its effects, was a sense of unreality, of helplessness, and anger. Because they were women, they could not go out and shoot a gun or punch somebody in the nose or get drunk. They had to hold their feelings inside.

Shad Meshad,
Army psychologist and Vietnam veteran,
Interview in The New York Times

Don't tell me women don't know anything about war because we weren't out on the frontlines. I had battle fatigue too, from those gruelling years in surgery. It was a war zone there, believe me. There were days when the stress and strain and blood and guts almost had to equal what you experienced out on the frontlines.

Juddy Marron,
Vietnam veteran, Letter to the Vet Reporter

On a hot afternoon in August 1984 Maggi Arriola took her eight-year-old son and set off to attend the unveiling of a statue of a Vietnam army nurse in St Paul, Minnesota. She was apprehensive as she drove to the ceremony. "I said to my son, 'This might be like Memorial Day. I might cry, and I want you to know that I'm not crying about anything that you've done, or because I am hurt or anything like that. It's an okay time to cry, so if I do, don't worry about it.'

"When we arrived the entire Landmark Center was full of people. They didn't even have enough chairs. After some speeches, they asked those nurses who had been in Vietnam to come up on the stage. I was a little hesitant, but my son kept pushing me and finally I went up there. There was a feeling . . . we were all standing there, and I wanted to put my arms around the woman standing next to me, yet I couldn't do it. I don't know why. I wish I had done it. You could feel the closeness in the air. The veterans in the audience stood up first, and then everybody stood up and applauded. It's not that I had been waiting for someone to say thank you, but nobody ever had. I was crying.

"When I got back to my seat, my son said, 'Mom, I feel so proud. My heart feels *this* big.'"

The thirty-three-inch bronze statue depicting an American nurse in Vietnam was commissioned by two former army nurses who had served in Vietnam, Diane Carlson Evans and Donna-Marie Boulay. Together they succeeded in raising funds to complete the statue, and the unveiling was attended by nearly forty former Vietnam army nurses.

The six women in this chapter all came together because of the statue. Several of them had not "come out" before, never talked about or acknowledged their service in Vietnam. They are now involved in raising funds for a larger-than-life version of the statue to place in Washington.

No one knows how many American women served in Vietnam. They were there as technicians, intelligence operators, air-flight ground controllers and WACs. Civilian women served in large numbers with AID, the Agency for International Development, the Red Cross and a number of other organizations. Most of the military women were army nurses. Even their number is in doubt. The Defense Department claims there were 7,465 women in Vietnam, 6,250 of them army nurses. Other estimates range as high as 55,000.

Seven army nurses are known to have died in Vietnam. Their names are on the Memorial Wall in Washington. Others have committed suicide since, unable to live with the memories.

There have been American nurses in all American wars. Only now are they beginning to receive recognition. When Donna-

Marie Boulay approached women veterans from World War II and Korea to support the Vietnam statue she was told that "no one had ever done that for *us*." Her response was, "No one is doing it for us, either. We are doing it for ourselves." Still, nursing in Vietnam was not like nursing in previous wars. Because of the nature of the war, of air mobility, many more severely injured soldiers reached the evacuation hospitals than ever would have in the past. It meant that more permanently disabled went home. It was different, too, in that it was a war in which the veterans were to become as unpopular as the military effort, blamed by the hawks for losing it and by the doves for fighting it. For women, there were additional biases.

In 1979 a former army nurse, Lynda Van Devanter, was asked to form a women's committee for the Vietnam Veterans of America. Under her direction, for the first time Post-Traumatic Stress Disorder and health hazards caused by herbicides and chemicals were examined for their effect on women veterans. In 1983 Lynda published *Home Before Morning*,[1] her own story of a year's "tour" in Vietnam. All the women I interviewed who had served in Vietnam were grateful for Lynda's pioneering. In part because of her, and partially because of the healing passage of time, more women veterans are now speaking out. Lynda's book records only one experience; each individual's twelve-month tour through hell was unique.

Like Geneviève de Galard, women in the second generation to go to Vietnam, a distant country they scarcely knew where to locate on a map, were motivated by patriotism or the simple belief that their skills were needed to save lives. The stories of the women in this chapter begin in 1967, a full twelve years after Geneviève's final days in Vietnam. During the French Indochina War, the Viet Minh had grown from a ragtag band of guerrillas to become, in the words of Colonel Bigeard, "the greatest infantry in the world." In the total fight they waged against their adversary, women served in every capacity, including combat. Ho Chi Minh would call them "the long-haired army."[2] It was this commitment to their struggle, in part, which defeated the French.

When the nationwide elections to unify Vietnam guaranteed in the peace agreements with France were not permitted by the

South Vietnamese government and their American ally, communist guerrillas in the South and regular troops in the North became a growing threat to a series of weak and corrupt South Vietnamese regimes. As the threat grew, so too did American economic and military aid to South Vietnam, including the services of American military advisers. The following years saw the assassinations of South Vietnamese President Ngo Dinh Diem and, three weeks later, of American President John F. Kennedy; and the coming of age of a generation of Americans who called themselves the Kennedy generation, patriotic and willing to ask, "not what their country could do for them, but what they could do for their country." Finally, those years saw the first American fighting troops land at Danang on 8 March 1965.

"In February 1966 my aunt's paper-boy came home from Vietnam with a portion of his head missing. He'd joined the marines and they'd sent him over there. He was only nineteen, and only there for a short time before he was severely wounded. I only knew him as Stanley the paper-boy, but that made a strong impression on me. All of a sudden I knew where my talents were most needed. I went to the army recruiters in Boston and said, 'I am a nurse, and I want to be sent to Vietnam.' The recruiter said, 'Raise your right hand and sign on the dotted line.' And then, 'What do you want to do that for?'"

Donna-Marie Boulay was from a small town north of Boston. She had spent the past three years nursing, specializing in trauma, while putting herself through college. At twenty-four she already had strong feelings about the Vietnam war before joining the service. "It was the time of life to be active," she says of the Kennedy generation she was part of. "I participated in the Medical Human Rights Committee in our college. Everybody was doing that kind of thing and thinking about what was going on in the strange-sounding country on the other side of the world. I was very politically opposed to the war, but when I weighed my political convictions against my response to my aunt's paper-boy . . . I knew I had to go. I knew I was needed there to help boys like him."

Like every nurse who signed up for the army, Donna-Marie

spent six weeks in basic training at Fort Sam Houston, Texas. The nurses practised drill, put up tents, simulated emergency room treatment and learned the fundamentals of triage, the procedure for deciding the order in which to treat battle casualties. "Then they sent us to the firing range one day and asked us to fire .45 pistols, the weapons issued to all officers. I told the major in charge that I wouldn't pick up a gun. I said I was a nurse. I absolutely refused. I will have nothing to do with guns."

When orders come through in the army, things happen very quickly. Although Donna-Marie asked to be sent to Vietnam, she nevertheless experienced a certain amount of shock and fear when she actually received her orders to go. "I was scared out of my mind at the thought of going over there. I don't think I ever overcame that. I said goodbye to my family with a lot of tears. It was helpful to know that I would only be there for a year." It was February 1967. The largest American offensive of the war thus far, Operation Junction City, was taking place only seventy miles from Saigon. American casualties for 1966 had just been announced at 5,008 men killed. The war was heating up.

Betty Stahl Doebbeling was nursing in a Minneapolis hospital emergency room. Also twenty-four, she was experiencing what nurses now refer to as "burn-out." Shabby treatment from patients who came in in abusive moods was changing her personality. "I was a happy-go-lucky person who was gradually becoming tight-jawed and unhappy. I was becoming irritable towards my patients, which upset me a lot. I got home one night and looked through my mail and there was a brochure from USAID [U.S. Agency for International Development] that said, 'Nurses needed in Vietnam' on the cover. I had become a nurse to help people. Now I thought nursing was making me start not to like people, and that was not my nature, I wanted to do something different, so I called the number on the brochure. Nobody I knew had been to Vietnam. I was totally uninvolved with the war. It was not as controversial then, and I hadn't really thought much about it."

Betty received information from AID immediately after her phone call. The huge civilian organization was operating in

South Vietnam on the principle of "winning hearts and minds" with economic aid and advisers. Hundreds of AID employees were sent to South Vietnam to advise the government there on everything from agriculture and economics to health care and garbage disposal. The most noted of its many programs to aid its uneasy ally was "pacification," in which South Vietnamese villagers were herded into camps away from their homes and ancestral burial grounds. These camps, or "pacification hamlets," were surrounded by barbed wire and guarded against attacks from the South Vietnamese guerrillas, the Viet Cong.

Betty: "They indicated that we would be teaching the Vietnamese nurses new techniques, helping the Vietnamese whose villages had been hit or who had diseases. I thought that would be great. We were told we would learn Vietnamese. That was really exciting."

A few months later Betty, who had never travelled farther from home than the neighbouring state of Wisconsin, was en route to South Vietnam. "My family was wonderful, very supportive. They took me to the airport, and that was the first time I realized that maybe I might not come home again. I wanted so badly to run off that plane. I looked out the window and I knew they were standing behind the long glass, and I asked myself what I was doing. I thought, 'I'm all by myself. I don't know a single soul. I must be crazy.'

"The airplane trip was hideous. We were in this great big Pan Am plane, and our knees had been in our chests for over fourteen hours. The pilot suddenly turned the nose down and we were spiralling, going down almost vertically. All I could think of was that we were going to crash, and nobody was telling us. We fell and fell and fell, and I thought, 'When are we going to hit?' I had to be soon because we'd been falling for three or four minutes. All of a sudden the plane broke through murky brown clouds and swung into a horizontal position. We were on the ground. Then the pilot came over the horn and said, 'Sorry about that folks, but they've got machine guns at the end of these runways, and this is the only way we can get in here.'"

The first sensation in Vietnam was the wall of heat that hit as the plane doors opened. Donna-Marie: "It was oppressively hot. The minute you started walking down the steps off the plane it was like a wall of heat that was going to push you over. I remember I had butterflies in my stomach."

Betty, who arrived several months later in June 1967, says: "They made us sit inside the plane for about an hour. It was so hot I thought I was going to die. Then they opened the doors and let us out and it was even hotter outside. I was plain sickly for about the next three months. I even contemplated coming home. I got everything. I was nauseated all the time, and every bug in Vietnam bit me. The bites would swell up so much that for a while I couldn't even wear shoes. I was miserable. I felt kind of sick all over from all those bites. It was very hard to sleep. It was such a wet heat that the sheets were wet, and you got chilly and hot at the same time."

There was little or no time to adjust. Donna-Marie was given orders to report to the 36th Evac Hospital in Vung Tau, on the coast of the South China sea, a former French resort town still being used for R&R (rest and recuperation). It was as far from the war as you could get in Vietnam, in an area supervised by Australian troops. (U.S. Pacific allies South Korea, Thailand, Australia and New Zealand committed a total of 7,250 troops to Vietnam in 1965.)

Donna-Marie was put up in a former French hotel in a comfortable room with a ceiling fan and the luxury of her own shower. Later, when she was transferred to the 93rd Evac in Longbinh, she would think back on her stay at Vung Tau as a relatively quiet and tranquil time, her accommodations there as "luxurious." It was only the first of many such contrasting images of this war.

American troops had been fighting in Vietnam since March 1965. Under General William Westmoreland, the Commander of American Military Forces in Vietnam, the strategy for taking the war to the enemy was defined. He deemed it time to take the war on the offensive, to search out and destroy the elusive guerrillas and North Vietnamese regulars even if it meant a scorched-earth policy. American troops, with their superior equipment and air mobility, were designated offensive forces. The Army of South Vietnam (ARVN) was used to protect military bases, cities and villages. In the end, critics would claim that this two-fisted strategy emasculated ARVN soldiers and diminished their will to fight, while it angered American GIs who suffered heavy losses and complained that they were doing most of the real fighting.

American military medical operations were set up to serve a given division or fire base. The hospital staff were usually notified before any major actions were undertaken and told when to expect mass casualties. In "dust-off" operations huge medical evacuation helicopters flew into combat zones, taking out the wounded and their buddies, and sometimes any and all survivors. Army hospitals had chopper pads next to their emergency rooms where triage was done to evaluate which patients would be operated on first, and sometimes which patients were too severely wounded to receive precious hours of surgery time. The objective of the medical facilities was to get the men operated on and stabilized, then evacuated as soon as possible to Danang or Japan for more sophisticated surgery and extensive care. Doctors and nurses in Vietnam who treated soldiers in trauma rarely knew whether they later lived or died.

Donna-Marie experienced triage for the first time after she had been at Vung Tau for a few weeks. "They had gone out on an offensive. Only twelve of the whole company [about 150 men] were left alive. It was the most gory thing I had ever seen in my life. Only two of the twelve people they brought in by chopper were still alive. I had to determine who was alive and who was dead. The young women who were with me really panicked and were convinced that all these dead people were alive — people whose whole necks had been torn apart or who were missing most of their heads. They were convinced that they were still alive. I remember shaking violently. My hands were shaking as I held the pulses and assessed the abdominal wounds. I was trying very hard to be as thoughtful and as carefully evaluative as I could. I was trying to exclude from my mind all the dead boys and the awfulness of how they had died and concentrate on the two living ones, figuring out which IVs to put in. I needed to be reassured on a minute-by-minute basis that the surgeons were coming, because the surgeons were right in the middle of OR and I knew I needed their help. I'm sure it only took five or ten minutes before they arrived, but five or ten minutes of that kind of hell is a long time. An awfully long time.

"Nothing could have prepared me for that induction into triage."

Betty Stahl was sent to work for USAID at the Vietnamese civilian hospital in Can Tho, a city in the Mekong delta. On 27 August a Viet Cong attack on the city killed forty-six civilians in a mortar-and-rocket barrage. Two hundred and twenty-two people, including five American soldiers, were wounded. "There had been some VC attacks on a few of the buildings we were in, and they were sandbagged. There was an attack on the hotel they put us up in just two days before we got there." Life consisted of enormous contrasts. The USAID personnel were put up in their own apartments with modern amenities, while the Vietnamese hospital to which they were assigned was overcrowded and filthy. Rats ran over their feet as the nurses worked, and patients slept four to a cot.

"The idea of training Vietnamese nurses kind of fell by the wayside because our Vietnamese wasn't good enough, and also because I believe we were quite offensive to them in some ways. I don't think the proper groundwork had been laid. When you come in and try to say, 'Our way is better,' without being able to say why . . . it doesn't work. I remember trying to explain things like scabies to them. When I told them they were mites under the skin, they would just laugh up a storm because that was impossible. Finally, we gave up teaching and just took care of the people."

Because of numerous attacks on Can Tho, USAID nurses were not permitted to work at the hospital after dark, when the Viet Cong usually attacked. At night the hospital functioned with a skeletal staff.

There was no time to get used to the war.

"The Viet Cong would attack a village and we would have casualties pouring in day and night, all night. One time our hospital was hit at two o'clock in the morning and mortars came in all around our area of town. They came and got us in truckloads and jeeps, still in our curlers, and took us to the hospital. I recall spending that night and the next day working . . . two days straight without sleep. One of the mortars had hit the water mains so there was no fresh water, just a couple of basins that the surgeons dipped their hands into, but the water was red with blood. The surgeons were to the point of exhaustion. They had no X-ray because a mortar had come through our X-ray roof. Our X-ray technician's wife and family had been killed, but he came into the hospital and tried to help in any way he could.

"We just managed to triage people as best we could, and they were coming in in hoards. We had wall-to-wall people on the floor and very few IV stands left. In those days the IV bottles were glass, with a wire handle. I remember a young mother, probably about twenty-five. She had two little kids, a baby and one about four years old. The whole back of her head was gone, but she was still breathing somehow. She was on a green army litter that was just filthy, and we lifted her up and carried her into the room where we put the dying. One of us got an IV going in her, just because the Vietnamese believed IVs were some kind of miracle. There were no more IV stands, so the little boy, with his tiny little hand, stood there and held the bottle until the wire handle cut right through his hand. He never cried, he was just a little stoic gentleman. He stood there while his mother died."

Days have become indistinguishable in memory. Only the most gruesome or shocking remain intact as days, separate events. Besides those, Betty remembers generalities. In the hospital's small burn ward most of the casualties were children, often the victims of home fires rather than napalm or combat: "USAID brought in gasoline for the Vietnamese to use, but it was different from the oil or kerosene they had been using. It was more explosive. So kids would be fixing the cooking fire on their houseboats and it would blow up in their faces. Some of the casualties we got were little kids who'd been swimming in the river. We had boats with long motor handles, a shaft which had a propeller at the end. These little kids would be swimming, and their beautiful long hair would get caught in the propeller and scalp them.

"I saw tetanus and diphtheria in tiny babies. We only had one suction machine, and it was filthy. Mothers would sit in a row and hold their babies whose throats were closing up on them, and they would take that one suction machine and go from one to the next, just to keep them breathing.

"It took getting used to, after the hygiene of American hospitals. I had never seen maggots used for a treatment before. The doctors left them in the wounds to eat the necrotic tissue. They looked like grey jelly that continually moved. It was very itchy. The doctors would keep a close eye on them, and when the bad tissue was gone they'd clean off the maggots and do the repairs."

As Americans and as women, the nurses received a lot of attention in off-hours. Betty, a tall blonde, was nicknamed "Princess" by the Vietnamese nurses she worked with. Men did not miss a new face, particularly a good-looking one. "I got a lot of attention," she says. "But I learned to bounce with it. Almost all of the guys were married, even though they would tell you they were single. I would always check the records if I was interested in someone. We had a lot of interaction with the military, a lot of dates, and a lot of fun.

"Many people drank a lot. Some of the older women were real alcoholics. I didn't feel at ease with that. It got very boring talking to people who were smashed all the time. They'd start drinking the minute their shift was over. I thought, what a waste. But it was their way of coping."

Donna-Marie had the same reaction. "There were always parties, and always alcohol, but I didn't do a lot of drinking because I knew I had to get up and go to work. When I got off I was exhausted. I almost always slept for a while. On days off we went to the beach. I dated a lot. Being young and single, I had a lot of boyfriends."

She found Vietnamese culture intriguing. "We had a professor from the University of Saigon come and teach us the language. I fell in love with the people, the culture, the cuisine. I used to go to the Catholic church to hear mass, which they still said in Latin. I worked in a local clinic on my day off and dealt mostly with women and children. They came as much to look at us as they did for health care. They were very curious about Caucasians. They would come up and touch you, and hear what your voice sounded like."

Betty had a very good ear for Vietnamese and could understand a great deal of it. "I'd have all these little kids on their cots all around. Their mothers would come in, and they would all sit together on the cots and look at me and discuss me. They didn't know that I could understand what they said. They would say, 'I wonder how old she is? I wonder if she's married? Do you suppose she has any kids?' I would always start laughing when they said, 'I wonder how much money she makes?' Then they would see that I was laughing and I'd tell them I could understand what they were saying. Then they would just swarm around me and ask me zillions of questions about my family, and invariably ask me how much money I made. I would say, 'Americans never tell!' We became really good friends."

Betty had been looking forward to the Vietnamese New Year, Tet, which was celebrated on 30 January. It was a time of feasting and fireworks, and a one-day truce in the long war. "One of the guys from USAID and I went into town on his motorcycle, to see all the flowers in the marketplace. We went down by the riverside to the market and looked at these georgeous baskets of flowers. But the crowd was different. It felt as if they knew something that we didn't know. It was very subdued for a holiday. The Vietnamese all avoided us. All of a sudden I got this terrible pain in my leg — a kid had come up and kicked me as hard as he could. I don't know what he had on his foot, but his kick rammed my leg into the motorbike. It was the first hostility I experienced directed at us. We were well known from working in the hospital, and normally people were friendly. I said to my friend, 'I just got kicked really bad. Let's get out of here, something is wrong.'"

Donna-Marie had also been looking forward to the Tet celebrations. She was now stationed at the 93rd Evac in Longbinh, only twelve miles northwest of Saigon. It was a far cry from the resort town at Vung Tau, and the casualties were more severe and more numerous. The evacuation hospital consisted of several metal Quonset huts with rounded roofs, joined together by metal bars. The ER was located within fifty yards of the helipad and connected to the OR cubicles, post-op, the X-ray room and the morgue. The compound was surrounded by barbed wire and guarded. Donna-Marie's accommodation was less luxurious here. Like all nurses, she had her own room, or hooch, which was large enough for a cot, footlocker and chair. Her hooch was located next to the bunker.

After work on the eve of Tet, Donna-Marie had a date to watch the fireworks display from the vantage point of the dust-off helipad. "After we watched the fireworks we went back and had a nightcap at the pilots' bar. We sang and closed the bar down at one o'clock, and then about four of us sat outside for about another hour. I finally got to bed just before three in the morning.

"I was in bed when the firing started. I thought Armageddon had started. Viet Cong troops were shelling the area. There were huge explosions. I had heard outgoing rounds many times, and the sound of incoming had been described to me. I put two and two together and knew something strange was happening. We were all evacuated to the bunker immediately. It was a depression in the earth with sandbags piled all around, and a piece of corrugated metal held down by sandbags for a roof.

"I was scared out of my mind. We stayed there all night. During those hours my period started. I was too afraid to go about six feet out of the bunker to where the bathroom was."

In Can Tho the light of flares woke Betty at two in the morning. "It was just like daylight, and at first I thought I had overslept. I looked at my clock and it said two o'clock. I thought, 'What in the world is going on?' I got out of bed and went outside onto the balcony. All my friends and people from USAID who lived in that apartment building were out there, lined up along the rail, looking at the sky. There were probably fifty flares out there. No one knew what was happening. It was deathly quiet. Not even a motor vehicle, and that was very unusual.

"Someone came and told us to go to the compound across the street, where we would be safer. We grabbed a few clothes and went. They bunked us all over the place. It was a real eery night. We slept in our clothes and waited to find out what was going on.

"Nothing happened so the next morning we returned to our apartments. A couple of Air America men were sent to guard us. All our Vietnamese guards had disappeared. By then Armed Forces Radio was reporting that the American Embassy in Saigon was under attack, and that the Viet Cong had taken over most of Hue. We slept very little for the next few days."

Donna-Marie left the bunker to report to work at seven that morning for her normal twelve-hour shift. The hospital was full of new casualties. "It was very heavy for us. We would evacuate casualties that we had just patched up and send them to Japan as fast as possible so that we could have empty beds for the people who were coming in. We could hear guns firing most of the time. It is a very tiring experience, and you kind of give up being afraid. I think it is kind of like a muscle that is overextended: before it can contract again it has got to relax. I think most of us were just in that overextended phase, and we were not given the opportunity to relax and contract again. I worked most nights after Tet began. I remember I could never go to sleep right afterwards. I would wait till after lunch. Lunch was the only meal I could bear to eat."

The Tet Offensive was launched by approximately eighty thousand Viet Cong and North Vietnamese soldiers. They simultaneously attacked South Vietnamese and U.S. army bases and most major cities in South Vietnam. Intense fighting lasted throughout most of the next month. In the United States the Tet Offensive, more than any other factor, heated up controversy about the war. The fact that the Communist forces could launch such a large and organized attack on the very heart of the South, particularly the American Embassy, was shattering news to those who believed the United States was winning the war. As casualties mounted, editorials more and more frequently questioned the price of the war. For the first time since the United States had become involved, venerated newscasters like Walter Cronkite publicly questioned American motives and methods in Vietnam.

Betty: "It was the most terrifying thing that has ever happened to me. I was terrified when they told us that we could not escape by river because the Viet Cong were so thick on the riverside. Then they said that the American airport which was about three miles down the highway from us had been overrun and almost everyone had been killed. Some guys who had been getting on a plane to go home had been shot right on the stairway of the plane. We couldn't escape by plane or by river, and the roads were absolutely off-limits.

"People were reacting very calmly. It is very strange. I never saw anybody get hysterical. There was just too much to comprehend at the time. I can remember going up to the rooftop when we thought things were quieter and asking the Air America guard if he had seen any Viet Cong. He said, 'There's one,' and I put my head up and saw one running around the corner, all dressed in black. The guard fired the rifle and said, 'I got him.' I felt sick. How did he know he was really Viet Cong?

"The next time I tried to get out of my room, somebody shot at me and hit the cement along the door right above my head. I slammed the door and dropped to the ground, then reached my hand up and tried to open it again. As soon as the door started opening they started shooting at me again. I couldn't get out of my apartment for three days. I couldn't even tell anybody — I didn't know where anybody was. I had no way of communicating with anybody. I had enough food, so I was okay.

"Those three days were really bad. The time is kind of lost to me. I know I sat at my little counter and wrote a letter to my parents, telling them the situation I was in, and the possibility that this might be my last letter. I wasn't even thinking how that letter would get out of there. I wasn't thinking rationally.

"At the end of three days the American gunships came in. They flew in so low that if I'd had a basketball I could have hit the underside of them. The noise was unreal. They were firing down right behind our apartment building. They must have got most of the Viet Cong because the next day we were taken to work again."

Donna-Marie: "The first morning after the offensive began, I was working and our supervisor was with me counting empty beds, figuring how many beds we needed with all the wounded expected in. We had already got our first Tet casualty, who was in the recovery room, and we were standing at the foot of his bed when the ammo dump that was about half a mile from the hospital blew up. The shock impact caved in the wall of our Quonset hut. This tiny little lieutenant colonel was shorter and thinner than I am (and I am only five feet five inches and 110 pounds). She had been in the Korean War and knew what to do. She looked at me and said, 'We've got to lift this patient and put him underneath the bed.' He was in a body cast. She and I picked this guy up. I don't know where in God's name we got the strength to do that. It would take four or five men. We got this huge burst of adrenaline which enabled us to be that strong."

Donna-Marie worked in the recovery ward, watching over the men as they woke up after surgery. "Usually they woke up and were surprised to see an American woman standing there. They had been through terrible things. Sometimes they were withdrawn. The thing that impressed me, and impresses me to this day, was the courage that these youngsters showed so often. It was real courage, not that macho thing when you've got a gun in your hand. I remember specifically one boy who'd been a mechanic. When he woke up he was missing both of his arms — all he had was maybe seven or eight inches of stumps. He got up as soon as he could and he walked around, pushing his IV pole around with just his two stumps, and he showed tremendous courage. He was good-natured. We all thought he was in shock, but after a couple of days went by we were convinced that he wasn't in shock, but that he had a magnificent personality. It was real courage, and that kind of rubbed off. It was awfully hard to maintain your own courage, but when you saw people like him they would really inspire you."

Though American forces soon quelled the Tet Offensive, that month accounted for the heaviest weekly death toll of Americans during the war: in total it took twenty-five hundred American lives. Many towns, including Can Tho, were badly hit by American airstrikes, which routed out the Communist forces but left thousands of people wounded and homeless. In all, half a million Vietnamese civilians were left homeless. For Betty it was the beginning of a realization that the American presence in South Vietnam was not always welcomed or appreciated. "I knew youngsters who for fifty cents would throw a grenade inside an American jeep. It began to get to me. You like to think that everybody likes you and everybody is glad that you are there, that you are a help to them, doing them a big favour. It really hurt when they showed hostility. I guess it was too overwhelming for me to even consider trying to figure out whether we should be there. The only goal I could work towards that would keep me going was to do as much as I could in the time I was there. Stopping the war was beyond me."

As Donna-Marie learned in Vung Tau, working predominantly with lightly wounded men had its own stresses; after she had helped patch them up, they would be sent back to fight. While under her care they were conscious and she got to know them. Sometimes they spent two weeks recovering on her ward. She remembers them fondly. "I was like their mother, and they were like my kids." She gave them tasks to keep them busy. "I'd turn them into instant nurses, instant corpsmen. They wrote letters for each other and helped push each other around in wheelchairs. We had music on. Sometimes we had parties." One patient was a nineteen-year-old boy named Randy Brown. The first time Donna-Marie treated him he was lightly wounded in one leg, hit by fragments of a mine that had killed or maimed most of his buddies, and when he was well enough he was sent back to fight. Donna-Marie remembers his enthusiasm and the comedy routines he put on to cheer up the other patients. The second time she saw him was two months later. He came in with a bullet wound that had ripped through most of the flesh in his left arm. It was a serious wound, but the surgeons managed to find enough flesh to sew it up and keep the arm operative. Again, Randy entertained the others as soon as he was able. Donna-Marie saw him again when he came in to visit some of his buddies who were under her care. He brought cigarettes, candies and a radio. He visited men he didn't even know an spent a while talking to her.

At Longbinh, during Tet, she was assigned to intensive care. By mid-February the hospital was largely empty, the patients having been evacuated to Japan to make room for large numbers of wounded who were expected shortly. Donna-Marie reported to work one night: "The day nurse and I sat down at the desk. She said I only had one patient in my section of the ward. He was in unusually bad shape. His unit had been attacked, and he had been hit by several automatic rifle bullets which had shattered most of his lungs and intestines. Grenade fragments had lodged in his liver and pancreas, and the surgeons were unable to remove them. One kidney was gone and the other one was shutting down. He had been unconscious since he had come out of surgery. He was too unstable to send on to Tokyo, and she doubted this twenty-year-old youngster would last the night. As I got up to go see him, I looked for his name on the chart — Randy Brown.

"I rushed down to the bedside. His skin was sickly yellow, he was scarcely breathing, and what urine there was in the containers was brown. He felt hot and dry, and his pulse was weak. I picked up his hand and just stared. He was almost unrecognizable, but it was Randy Brown."

"'Randy,' I whispered, 'Oh, Randy.' It was like discovering one of my children lying in my intensive care unit. I took his temperature — 105 degrees. His blood pressure was inaudible, so I started another unit of blood. I kept talking to him. 'Randy, this is Miss Boulay. Wake up, wake up please.' I gave him his medication and rubbed his back. 'Randy.' He was still breathing. I kept on talking, 'Randy.' I changed his bandages. He was bleeding profusely from the massive wound in his abdomen and from where his kidney had been. 'Randy.' 'Hi, Miss Boulay.' 'Hi, Randy.' We talked slowly, softly — no jokes, no smiles. He wondered where his leather medallion was. I held his hand. His eyes were sunken, grey and yellow.

"About midnight he closed his eyes, his breathing shallow. We stopped talking. I kept holding his hand. Soon Randy Brown died. Then I cried."[3]

Two weeks later Donna-Marie's tour came to an end. She was not sorry, she says. "I can't even remember too many people from the 93rd. We couldn't get close. It was all intensive care and recovery room work. I remember that I was supposed to leave on a certain date when my year was up, but they couldn't get safe transportation for us down to Saigon. We had to wait a few days. I remember getting to Saigon and borrowing a lieutenant's room to change in. I took off my fatigues and boots and shoved them under his bed, and left him a note to throw them out with the garbage in the morning. I didn't want to bring back anything to do with the awful misery they symbolized in my mind."

She stayed in San Francisco and then went to work at a

hospital in Tacoma before returning to see her family. She felt she was not ready to face them when she first got back. "I just couldn't go home and explain what I'd been through. I couldn't share all that misery. I was so afraid I would cry and they would think I was terribly traumatized. I'd call them and talk to them, but they let me set the pace for whatever I wanted to do or say. I underestimated them. I should have gone home immediately, but I couldn't. I needed to get my fear and anger under control."

Tet had sent seismic shock waves through America. One hundred South Vietnamese villages, towns and cities had been simultaneously attacked, and the weakest defences had crumbled. An aghast Walter Cronkite exclaimed, "What the hell is going on? I thought we were *winning* the war!"

Most of the Tet attacks, however, were quelled within a week and Communist losses were extremely high. Once the shock subsided, American and ARVN forces subdued or pushed back their Communist attackers. Tet had not been a military defeat for the United States, but the real battle of Tet had been fought and lost in the heart of that country. The stalemate it proclaimed to all who watched the news each night at six had forced a deep and dreaded look into fifteen years of involvement and three years of fighting in Vietnam. That look forced all but one of President Johnson's advisers to turn against the war. It presented the stark reality of a political involvement without a viable strategy, a military involvement without a workable set of tactics, an ally without a strong enough will to fight. The real battle of Tet led to an outcry from Americans that they wanted the truth and, finally, wanted out.

On the evening of 31 March 1968 Donna-Marie was alone in the living-room of her nursing quarters in San Francisco when President Lyndon Johnson announced on nationwide television that he would not seek re-election. She says, "I'll remember that telecast as long as I live."

Richard M. Nixon was elected president in November of that year on a platform of "peace with honour in Vietnam." The idea was to gradually withdraw all American troops, turning over equipment and responsibility to ARVN. During 1969 another 9,141 Americans lost their lives in Vietnam. The controversy at home did not lessen. On campuses throughout the country young men chanted, "Hell, no! We won't go!"

It was a sorry home-coming for the veterans. They were spat at and called baby killers for having gone and fought a war that their government had forced them to fight. Families were divided between those whose conscience told them to serve their country and those who believed that the war was a mistake. When First Lieutenant William L. Calley Jr. was accused and later convicted of the murder of over a hundred Vietnamese civilians at the village of My Lai, the public was outraged by the atrocity. Still, it was divided outrage. Many felt Calley was a scapegoat for all the brutalities that the years of seemingly senseless strategy had wrought. No one condoned what Calley had done, but many people realized that My Lai was not an isolated incident.

Perhaps like many of the returning veterans, Betty Stahl felt confused and less in touch with herself than she had in Vietnam. Many veterans signed up for a second tour because they found it easier than dealing with being home, in a society that was often hostile. For all who served there, Vietnam was the most significant twelve months of their young lives: danger, friendships, love and anguish, even combat, seemed easier to understand than what life offered them in the United States. To some, Vietnam became an addiction. Betty left USAID after eighteen months in Vietnam, returning home in November 1968. After a few weeks she walked into a local army recruiting centre and signed up, asking for service in Vietnam. "I guess I missed it," she says. "I felt I had done all I could do for the Vietnamese, and now it was time to do something for my own country. It felt really empty when I got home. I missed the challenge. Maybe it was a kind of addiction. Nothing else seemed as important."

Assigned to the 91st Evac at Chu Lai, a northern city on the South China Sea, Betty found her quarters much more austere than her USAID apartment, but the hospital was "a luxury. There were supplies in the cupboards, medicine and bottles of blood. The floors were clean." Furthermore she was glad to have patients who spoke English and get away from caring for children. The child victims of the war left the deepest mark on her. Today she says she can work in any kind of nursing except paediatrics. Dealing with American administration was easier. She had very hostile feelings towards the Vietnamese officials, whose corruption was so deep-rooted and difficult to get past that of the shiploads of medical supplies USAID consistently sent to Vietnam, only minuscule amounts arrived in the hospitals. She says, "It was a relief not to have to beg for each bandage, as I had had to do at AID. The Vietnamese doctors were getting rich, living in Paris, while we fought their war. That was what made me the angriest — the corruption. The Vietnamese government even charged the Americans *rent* for the land on which we built our hospitals. And we were shedding our own blood for them. Rent! I was so incensed. I thought, 'If they don't want us here and don't want to take our help the way we can give it, without making us *pay* for it, something is terribly wrong and they don't deserve it.'"

There were a lot of acute differences to get used to. Betty was assigned to intensive care. "I had twenty-five of the more severe casualties on one side, and the lesser ones on the other. Every ward had a stereo, and I can't stand loud rock music. It would be going on all the time. I wonder how many of them liked it. I wonder how much was imposed on these men who weren't feeling well anyway. I had to spend a couple of nights in the recovery room after I got a little blood clot in my leg operated on. I can remember those tapes being played. One in particular offended me beyond words, and it was played over and over again: it was "Walk Like A Man," and many of these boys were never going to walk again. I thought, how cruel, to play something like that. I know they didn't mean it to be cruel, but I thought, 'Use your heads when you pick out those tapes.'"

Betty was by now a seasoned veteran, and at twenty-eight older than most of the nurses at Chu Lai. She was made supervisor and became something of a mother hen to the other nurses. So it was she who first met and befriended Sharon Lane, a twenty-six-year-old nurse from Canton, Ohio. She remembers Sharon as a lonely girl whom she tried to introduce to others. One Sunday the two women were talking, and Sharon revealed that her boyfriend in the States had a terminal disease. "She added, 'And I'm not going home either.' I said, 'You don't know that, don't talk that way,' but she said, 'I know I'm not going home. I said goodbye to my parents.' I was trying to comfort her and not make fun of her. I thought if she was living with that kind of fear, it must be terrible for her."

On 8 June 1969 at a quarter to seven in the morning, the hospital at Chu Lai was attacked. Betty was getting ready to go to work when the first mortar hit: "I had one boot on and was lacing it up when the mortar hit. I don't know how many there were, because after the first one you are only worried about covering your head and your body. I flew under my bed and put my helmet over my face and my flak jacket over me. Then I started having all these visions of spiders and vipers crawling under the bed. I heard plaster and glass and everything crashing, so I knew we were being hit really close. As soon as it got quiet I got up from under the bed.

"There were probably four or five mortars that hit us. One of them hit the Vietnamese ward, square on. Sharon had been sitting on an empty cot by the door waiting for the next shift to come it. The mortar came through the roof and knocked down the fluorescent lights on top of the beds. The water pipes were hit, so there was water and hot electricity all over the place, and glass everywhere. The nurses' desk had spun around and looked like Swiss cheese, it was just full of holes. A piece of something hit Sharon right in her jugular. She never knew what hit her. She apparently flew up in the air and landed on a bed with such force that it flipped over. When they found her an old Vietnamese and a six-month-old baby were lying on top of her.

Sharon Lane was killed instantly. Another nurse on the ward received a minor wound but was psychologically traumatized and was sent home shortly thereafter. Betty: "It made the camp very, very quiet for the next few days. I played the organ for her memorial service." Lieutenant Sharon Lane was the first U.S. army nurse to be killed by hostile fire. Several others had died in helicopter crashes during the war, and many more were wounded.

Three young men came home from Vietnam about the same time after the Tet Offensive. They each had a sister who was nursing. Mary, then a twenty-one-year-old nursing student in

Minneapolis, remembers her brother saying how badly nurses were needed in Vietnam. "Later, when I joined up, he was shocked," she says. "He said he hadn't meant me to take it personally." Lynn Bower, then only twenty, had sworn that if her brother didn't return from Vietnam, "I would never live in the United States again." He did come home, but was withdrawn and never talked about Vietnam. Maggi Arriola's brother came back and used his veteran's benefits to go to college. She remembers that he locked himself in his room on campus during the moratorium against the war. She did not have an opinion about whether the war was right or wrong, but she felt strongly that the protestors were hurting veterans like her brother.

Mary Beth Crowley did not have any brothers in Vietnam, but by 1969 almost everyone in the United States had been touched by the war. Like Mary, Lynn and Maggi, she joined the army as a nurse.

Not only did these four women have to overcome their own misgivings about the rights or wrongs of their country's involvement in Vietnam, because they felt that as nurses they were needed, they also had to overcome social pressures. If American society was hostile to the veterans who were coming home from the war, they had further disdain for women in the military, particularly those who went to Vietnam. "You were either a whore or a lesbian," says Mary. "Why else go half-way around the world with all those men?" Lynn was spat at and called a whore. Maggi was called a warmonger. When they arrived in Vietnam they were confronted with the new and final phase of American strategy — the "Vietnamization" of the war. It made little sense to the men who were still losing their lives there. They would see their friends killed in the taking of a hill that, once secured, would be turned over to the ARVN. A week later ARVN would have lost the position, and the same American men would be asked to go in and take the hill again. The American soldiers' frustration in the face of this self-defeating military policy led to brutality, drug abuse and "fragging" — the intentional killing of officers by their own troops. All these aspects of the war became a part of the daily events witnessed by the nurses and doctors in the ER.

Lynn Bower spent her first night in Vietnam at the 90th Replacement Center, where nurses went to receive their assignments. "I went over to the Officers' Club. I was sitting there and things were beginning to hit me, like what was I doing here? This drunk colonel came up to me and told me what I owed him. It was his last night in-country, and he had been over there fighting for God, country and so forth, and it was my duty to accommodate him that night.

"I had been fairly protected growing up, so I didn't know what to do. There were a couple of guys sitting at another table who heard all this. They came over to me and said, 'Where have you been? We've been looking all over for you.' And they just took me by the arm and led me out of there. It was so nice of them. I really appreciated that. Then I decided that I didn't think I was going to like being a woman over there."

Lynn was sent to the 93rd Evac at Longbinh, where Donna-Marie had served two years before. "The troops were pulling out and there were none between us and Saigon. It wasn't a very comforting thought, but it wasn't Tet. The first day I arrived I was told I was needed in the ER for triage. I went in. It was a big room near the chopper pad. There were things like saw-horses to put the litters on, and next to each one were IV stands.

"There was no time to adjust. They called 'litter' and I rushed to the stand, and oh my God . . . they brought him in. He was sandy haired, very fair skinned, with freckles. He looked like half the boys I knew in Minnesota. He looked so young. I started cutting off his fatigues and took his blood pressure and put the number on his forehead or his arm, I can't remember. I went to start the IV, and then someone said, 'Lieutenant, you can't start that IV, it isn't going to do him any good.' I said, 'I have to get it in.' He repeated, 'It's not going to do him any good. He's dead.' I couldn't see it. He was a bit rumpled, but he was in his fatigues. His eyes were shut and he was quiet, but . . . I had never seen anybody young die. Then they rolled him over and his whole backside was gone. I couldn't believe it. I remember the room started spinning. I was saying, 'No, no. I'll just get this IV in and he'll be all right.' He was just too young. I wanted to be sick. The captain came over and said, 'Let's put you at the desk for a while. You can do paperwork.' So I sat at the desk, feeling a little better. Later in the afternoon they brought in some guys who'd hit a land mine. Five of them were young, and they had their sergeant with them. He was crying. I had never seen anybody older cry.

"I was confused. He had been there for two tours, yet he was crying. Their faces were so dirty. I had never seen anybody really dirty in a hospital before. They had black camouflage on their faces, and I remember because his tears were coming through this black stuff.

"They brought them in on litters. Not one of them lived, except the sergeant. I was at the desk and I didn't have to do anything physical. I had to write the death pacts; there was a seventeen-year-old and a twenty-three-year-old, and it was his birthday that day. There was a nineteen-year-old who'd been set to go home the very next day. That was why they were coming in. They were taking him to Ben Hoa. He was going home.

"I was sitting there, and I got angrier and angrier. I thought, 'Who is going to tell their parents what happened to their children? They sent real people off to Vietnam, and they are getting back a body bag. They are so young, so young, God. What are they doing here?'"

Maggi Arriola received orders to go to the 91st Evac at Chu Lai, which Betty had only recently left. She was an experienced nurse who had spent three years working in an emergency room in Alaska, so she was assigned to ER. "Nothing could have prepared me for that," she says. "I was there about a week when a Chinook helicopter was shot down. We got about forty or fifty casualties at once, all in different states. That was when I had to come to grips with triage. I really had to deal with it. I always thought I was Supernurse, that I could help everybody, that as a nurse I *had* to take care of everybody. When I had to realize that I couldn't take care of everybody . . . it was very difficult. In civilian life, if we had a bad car accident we always took care of the worst person first, not the one with the most likelihood of being saved, which is how the triage setup worked. I had to rationalize that if I wasn't doing anything for one boy it was because I was doing more for someone else. I remember yelling across the room many times, 'Did somebody get the chaplain? Is somebody over there with him?' I couldn't handle the thought that a boy would die alone.

"I knew I couldn't let myself cry. I couldn't allow myself to do that because if I did then I was taking time that could have been better used to help somebody else. It was important for me not to cry in front of them because I was whole, and being whole I felt that by my presence and closeness I could help them a little bit. I'd give them a hug, trying to block out my own emotions. I always asked their names and called them by their first name. I thought that was important, particularly in the military where you are known by your last name.

"I remember one boy came in who had a traumatic amputation of both legs and arms . . . and both eyes were injured. I remember starting the IVs on him, and he asked if there was a rabbi there. I said, 'No, we don't have a rabbi, but one of our doctors is Jewish, would that help?' I sent someone to get him, and meanwhile this kid was getting really anxious, and I kept telling him that someone would be there in a minute. He said, 'I know I don't have any arms, and I know I don't have any legs, but just tell my mother that I love her.' I began crying, right there. That was the only time I can remember crying in the ER. I remember trying to start the IV and crying. The ER was full of people, and I was saying to myself, 'Maggi, you can think about this later, you can't think about it now. You've got to go on.' The Jewish doctor arrived and sat with him. I don't know if he lived or died, I don't want to know. You couldn't know. That is why I tried never to learn their last names."

Mary was sent up to Quang Tri, to the 18th Surgical Hospital, the northernmost American army hospital in Vietnam. "I was naïve," she says, "Boy, weren't we all. I was twenty-one years old. My brother had been in Vietnam, my father was a World War II vet, injured in the Philippines. I thought my country must be doing this for a positive reason. I learned quickly." Mary Beth Crowley was sent to An Khe, an outpost between Pleiku and the Central Highlands from which the 4th Division made pushes into Cambodia. It was an isolated hospital and dangerous. Many nurses and medical staff had been wounded. On her first night there, a nearby explosion knocked the mirror off her wall and shattered it — "not a promising start."

Maggi had been confused about the war before she arrived, not knowing whether she thought it was right or wrong. One day she asked the Vietnamese interpreter what the South Vietnamese would do when the war was over and the Americans were gone. "He looked at me strangely and then said, 'There has always been war in our country.' I had been trying to figure out why we were there, and what we were trying to do to justify all these horrible things I was witnessing. That is when I began to question it all. I didn't dwell on it, though. I couldn't, or else I would have been unable to cope for another eleven months."

Unlike the soldiers the women nurses had no release from the fear and anger. For them there were no prostitutes, no guns to shoot off. They couldn't be crazy. The sexual double standard followed them to Vietnam. Women, especially nurses, were expected not only to deal with their own emotion but also to nurture and comfort and understand the emotions of the men they treated and served with. By profession nurses were expected to deal with trauma. There was little or no psychiatric help for them in Vietnam. Indulging in normal releases, like sex, merely slandered the reputation of women in the army.

Mary Beth: "It really screwed up a lot of women over there. There were a lot of nurses who were not all that attractive, and being in a situation where you were one in fifteen thousand meant you could have your pick of men. Some of them really fell into that and believed everything that was told to them. All those promises, getting engaged . . . all those lies. I wanted to say, 'Come on, wake up! This is just for now. It's not for keeps.' So many girls got their hearts broken. I had dated a lot before I got to Nam and took everything with a grain of salt, so I believed every guy was married until proven otherwise. But those girls really did get hurt."

Mary: "The guys thought we were sluts. They didn't care what you looked like. I was always taught to respect myself and make my own decisions, so I was all right. Still, society was changing in the States. There was a sexual revolution going on, and in Vietnam I saw so many of my girlfriends used. Married men would give them this big line about going home and getting a divorce, when they were just using them during the war. Some of my friends have mental problems because of that. The guys would go home and you knew darn well they weren't going to write or stay in touch. But you couldn't say that to your friends who were so deeply in love. It is very understandable that women wanted and needed physical love to counteract all that they were seeing and going through. It was very hard not to. I put all my love into the local orphanage. I knew better than to fall for any man."

Both Maggi and Lynn's fiancés were in Vietnam, and when possible they spent time off together. Both say they experienced a lot less harassment because they were known to be engaged, but Lynn says there was a guard outside the nurses' ward who was there to protect them not from the enemy but from rape.

Mary Beth remembers socializing a lot at An Khe, which was a small MASH hospital in an isolated area, often under fire. She says the continual danger brought the staff very close. After working all night, she recalls picnics of weiners and beans and lots of rum and Coke. "I often slept with my clothes on and one foot on the floor to keep the room from spinning," she says. "I have never drunk that much before or since Vietnam. But we were all sober when we were working. I really want to stress that."

Vietnam meant many things. A first time away from home and a first exposure to another culture for some, first exposure to sex for many, first exposure to trauma and death. Isolated images and even sounds remain, like the rock music blasting from tape decks. For Mary Beth the song that was played endlessly was "Save the Country" by the Fifth Dimension. "We used to play it when we changed shifts at seven in the morning and seven at night. All the nurses and corpsmen would dance around the ward, and our Vietnamese patients thought we were crazy. They probably thought, 'No wonder these crazy Americans are losing the war!' But they got into it too and clapped their hands."

For Mary, Vietnam meant an almost brother-sister relationship with her corpsmen (male nurses). When they discovered her strong religious beliefs they all became extremely protective of her. When her father died, despite all the death around them they were very upset and collected money to help her get home. When the cooks discovered she was saving almost all her salary to pay off nursing school debts, they let her sneak through the meal lines without paying. Although she was only twenty-one, most who knew her called her "Mom."

Many nurses mentioned other nurses who made money in Vietnam. Mary Beth had numerous offers from prospective pimps. When Mary sent home substantial amounts of her pay cheque the pay officer insinuated that she "must be having a good time." When Lynn told her fiancé about a very popular nurse down the hall who "always had visitors," he laughed and said, "And they're all male right? Well, she is going home a very rich lady." The nurses who were not promiscuous suffered by affiliation — and those who merely had a "relationship" were labelled "sluts." One of Mary's best friends at Quang Tri had to leave when she discovered she was pregnant. "And my dear sweet little corpsman said to me, 'What a slut.' I just yelled at him, 'So, you never have sex? Don't give me that double standard.'"

Mary Beth discovered that wearing perfume and eye shadow cheered up her patients. "I kept writing home for Midnight Blue mascara and coloured hair ribbons," she laughs. "We cheered up those guys just by being there. Once they picked up on what kind of person you were, they didn't bother you. They were just happy to have a 'round-eyed' girl to talk to. Even if I was exhausted, I'd sit and talk with them for a while and let them tell me about the girl back home."

Lynn remembers a number of marriage proposals from wounded men: "I'd tell them I wasn't available. You wouldn't believe how insistent they were. They said if they died they wanted to be able to say they were married. Say it to who, I don't know. I think they wanted something to own, to possess — something they could say was theirs, in all that craziness."

Mary: "If you stress anything, please let it be the extreme youth of those soldiers. To me the most important thing about Vietnam is that they were kids. They should have been at home, in school, with their families and their girlfriends. They were just kids. Old men do not fight wars. These boys were so young that I got a mother image at twenty-one. They were just babies. They were so naïve, so willing to do anything for their country, absolutely anything. That war warped their minds.

"Do you know what wars do to nice little boys like your brother? They saw their buddies blown up and children used as booby traps. It literally warped their minds and how they looked at other human beings. Their hatred for the Vietnamese. . . . I took one boy with me to the orphanage one day, just before he was going back home. Afterwards he thanked me. He said, 'I'm not over it yet . . . but at least now I can look at these people and think that they are human.''

Maggi found it happening to herself. On top of caring for wounded GIs and Vietnamese civilians, the army nurse had to care for POWs. The first time she was confronted by a POW she felt a great deal of sympathy for the patient, a boy who was only about fifteen. But the next time she had spent nearly thirty-six hours on duty and was exhausted: ''Many of us did things we are not proud of. I was tired, I had seen all of these young American boys who had been blown away and now I was being asked to take care of this person who, while probably not directly responsible for what had happened, was a part of what had happened to all the boys I'd just been with. I know that I could probably have gotten the IV in on the first try, but I didn't. His wounds were not such that he needed a urinary catheter, but he got one. He didn't need a nasal gastric tube, but he got one. That was my way of reacting. That was the day I came to realization that some of the body parts I had found in the GIs' clothing, and that sort of brutality . . . that while I could never condone it, I could somehow emphathize with their frustrations. They'd seen their buddies die. I think if you weren't there, you can never understand that. I think we all tend to have the fear that if we told you everything you would judge us.''

Mary remembers feeling hostility towards a woman guerrilla she had to care for. It was difficult for many to keep those feelings out of their minds while nursing the enemy.

Drug abuse was becoming more common among the soldiers. Lynn remembers: ''This guy came into the ER shaking and sweating. He needed morphine, smack, heroin. Where was everybody? He came up to me; maybe I looked like a young kid to him and he thought he'd get some from me. But I didn't have anything; we didn't keep drugs in the ER. He said, 'Please, please, I need something.' He got down on his knees. 'I need something.' I wanted to put my hands around his neck and strangle him: why was he doing this to me? I couldn't help him.

There was nothing I could do. He was sent to a detoxification center — they took him away. I felt so helpless. Here I was in this helping profession, out of this nice school, and no one had told me that really not-nice things would happen."

There were always the conflicting images: Lynn was working in post-op the day Miss America and the USO came to her hospital. "They put on a little show outside the Quonset huts, in the circle area outside. Then she came through the wards to the guys who couldn't get out there. And Jesus, those guys would have gone through ten hoops for her if she'd asked them. They were so happy to see her and the USO girls, all dressed up, looking great. And then . . . they left. I looked down the row of beds and saw that several guys had hard-ons. I wanted to strangle her. The guys were so glad she came, but then she was gone and it was very quiet. Several of the guys had tears in their eyes. They were stuck with their emotions, and we were stuck with them."

Fragging, the killing of American officers by their own troops, happened frequently in Vietnam. It was another side of the war that was brought into the hospital wards. Mary: "It was so sick. The whole war was a bunch of political garbage. America is one of the most powerful nations in the world, and we could have won that war. Instead, the leaders played with people's lives. A lot of people have a lot of accounting to do. They'd send the boys up to take a hill, they would lose maybe seventy of their buddies taking that hill, and then they would be told to pull out again and not secure the area. They would turn it over to ARVN to secure. Maybe a week later, they would be told to go and secure the same hill — and lose more of their buddies. They didn't want to go. So they would label the lieutenant who gave the orders an SOB and guarantee that he would be killed.

"I remember my first exposure to fragging. The men had gone into Danang at night to get their lieutenant. They took a grenade and threw it in his sleeping quarters, but someone didn't evaluate very well who was in there, because one of the sergeants they loved was in there. There were six casualties. Two were DOA [dead on arrival]. The sergeant died, and the lieutenant was taken to intensive care.

"After I'd finished in OR I went to make the rounds to see how the guys were doing. Some of them noticed my OR hat and said, 'Whatever happened to the SOB so-and-so? I said he was in intensive care and his family had been notified. Someone said, 'I hope he dies.' I said, 'My God, what is going on here? I don't know what I came here for, when it's come to this, when Americans kill Americans.' The guys said, 'You don't know what happens out in the field.' And then one guy in the group said, 'You just came over here to get laid.' And I tell you, my corpsman just about went for his head. I hollered, 'You don't know me, you know nothing about me, and if that is what you think you are sick, and I don't think you are worth saving.' I walked out. My girlfriend who was nursing on the ward told me later it was quiet for fifteen minutes after.

"The lieutenant did die. . . .

"I cried a lot. I cried all the time. I think that has helped me today."

Maggi: "The one thing I wonder to this day is how those men are coping now. So many came back unable to live with the things that they did. [Of 2.7 million who served, 500,000 received other than honourable discharges.] The suicide rate for Vietnam vets is 25 percent higher than for others in the same age range. As for saving the very severely wounded? Sometimes I asked myself if we were doing them or their families any service at all by sending them back like that . . . the really badly disabled men."

Mary: "The only case I saw of someone given Valium and allowed to die — a boy had lost everything from the waist down. Everything. One arm was also gone, and one eye. Some people in triage never would have touched him. He would have been one of the ones you tried to let die comfortably. But whoever that doctor was, he decided to try. Another doctor walked into post-op after the surgery and said, 'My God, who decided to do this? You are going to let this kid wake up and see himself that way? I won't let him wake up and see this.' So we gave him Valium, and he woke up a little but never comprehended. We had covered him with sheets, so he couldn't see. He died within twenty-four hours. I told a friend this story years later, and his response was, 'Who has the right to decide life or death?' I got very nasty and said, 'Where the hell were you?'"

Over fifty-seven thousand American soldiers died in Vietnam. Another three hundred thousand came home permanently disabled. Estimates put the Vietnamese dead at two million. All the casualties will never be counted. Only today are American veterans beginning to receive some recognition of their sacrifices for a war, right or wrong, that they were asked to fight. Only now is the American government beginning to make up for the terrible home-comings endured by teenaged veterans: for the first time, with the unveiling of the Memorial Wall in Washington in 1982, it has thanked them.

All the women in this chapter suffered what is called Post-Traumatic Stress Disorder after their tours in Vietnam. Fear, anger and sadness come out in various ways. Mary still cries and still has nightmares. Most of all, she can't bear the sound of helicopters. Lynn still looks for something to crawl under when she hears sirens. She too has nightmares. Mary Beth woke up one morning six years ago in a state of severe depression. She cried for six months. Finally her mother said she thought it might be Vietnam. "'Vietnam, no, that was years ago,' I told her. She let me talk about it, and I realized that it probably was it." Betty and Donna-Marie have had longer to readjust, but both remember nightmares and weeping spells.

The Memorial Wall helped them get over some of the unresolved, buried anguish, the relentless images that stayed with each woman and could easily be flashed through the consciousness at any time, provoked by a smell, a touch, a comment or a song. When the Wall was built some of those images were finally laid to rest. Each of these women said something about the Wall. Some have been to see it; the others all say they plan to go. They talk about it as a pilgrimage — necessary and cleansing. Lynn said, "When the Wall went up I didn't have my dreams of their faces anymore. It was as though I didn't have to keep having my dreams because they were remembered. Somebody else took on the responsibility of remembering.

"All those faces of boys, simply boys. I don't think I saw many men die; I saw boys die. It was so futile. It was as though everyone had forgotten them. I had awful feelings of frustration and anger. When the Wall was built I was released from some of that. The Wall made them visible."

Nursing has always been an invisible profession. One thinks

of doctors when one contemplates illness, or of soldiers when one contemplates war. For the nurses who were in Vietnam, there were no thank-yous, no public recognition of them as veterans, no groups to help with therapy. They are entitled to veterans' benefits, but the bureaucracy is far from attuned to the needs of women veterans. Only now are the nurses beginning to "come out," to admit that they were in Vietnam and, when necessary, to seek therapy. Nurses undoubtedly saw the worst of the war and saw it relentlessly. Unlike the soldiers they never had "down time," weeks of inactivity during which losses could be mourned and spirits restored. They saw death, not just among unknown soldiers or civilians, but among friends too. They lost men they were dating, medics they were friends with, helicopter crews they had grown to love. Most of all, they dealt with an unbroken procession of wounded and dying boys. They are not like the rest of us.

Mary says that friends in the U.S. Reserves have recently had their nursing skills evaluated for a possible war in Central America. "My friend said, 'This war won't be like Vietnam. It will be different. In Vietnam you evacuated the guys to Japan. If there is a war in Central America, those kids are going to come right back to the States. People enjoying life here will have to see the horrors of war and relate them to human beings. It will end quicker.' My response was, 'Why is it going to start?' The truth of it is that people do not learn from history."

LASTING IMPRESSIONS
The Falklands, 1982

> We know from the success of these films [portraying war as a grand adventure, thrilling and exciting] that there is a tremendous amount about war which people really do love. I don't think they would love it if they saw it.
>
> ***Linda Kitson,***
> *War artist, the Falklands War*

For the journalist, photographer, artist or documentary film-maker there is no bigger story, no more compelling experience to record, than war. Today more women than ever are going to war in these capacities.

The first accredited woman war correspondent was Canadian journalist Kit Coleman, who covered the Spanish-American War of 1898 after bullying her way into the American Secretary of War's office and harassing him into signing her press card. While her Toronto newspaper, *The Globe and Empire,* ran the headline "Our Own Kit Goes to the War," she was being refused passage to Cuba by both military and medical ships. Finally she hitched a ride on a fishing boat and arrived in time to cover the surrender of the Spanish in Santiago.

Today there is a legacy of dedicated, courageous and insightful women war correspondents to inspire the young women who are making their way to San Salvador, Managua, Beirut and Iraq. Brigitte Friang, Marguerite Higgins, Frances Fitzgerald, Gloria Emerson, Catherine Le Roy and of course Oriana Fallaci are but a few whose words and images have helped us to interpret the events and circumstances of war in our time.

The experience of one woman observer, Linda Kitson, is unique. She was sent to record war rather than participate in it, but unlike a correspondent she was not asked to analyse it. When I heard about her I was immediately intrigued. She was such an unlikely individual in a combat zone, and yet this London artist and avowed pacifist was sent to the Falklands War as Britain's first official woman war artist.

It seems amazing that in the age of fast-paced reporting and television, artists are still sent to war. The Falklands War itself was striking: in the age of supersonic warcraft much of the fighting was done in trenches, with bayonets. It was a war in which Britain did not allow women to participate, and yet the country was led to war by a woman. Prime Minister Margaret Thatcher's actions, like those of Golda Meir and others, clearly indicated that female leadership is not necessarily more pacifist than male.

There are other striking ironies. The Falklands War, despite its political and geographic insignificance, aroused a patriotic, gung-ho, old-war mentality in Britain, a feeling that has not been prevalent since modern covert wars, Third World revolutions and the nuclear threat have obliterated front lines.

Finally, Linda herself is fascinating. As a young western woman with no previous exposure to the military, deeply committed to dialogue on arms control, she, like most of us, never expected to find herself in the midst of a war. Her confusion over her reactions to it strikes respondent chords. Conflicting emotions of pacifism and patriotism often run deep. There is a gripping immediacy to the lasting impressions she shares.

It is a cold December night, and we sit huddled over coffee. Tacked to the walls surrounding us are drawings whose stark, powerful images bring to life the almost Crimean look of the Falklands War. Among the drawings is a photograph of Linda Kitson looking diminutive in heavy combat gear.

"It's been a year and a half," she says, "and yet I still wake up and go to sleep with it. I suppose it is because I saw such extremes and felt such extremes of emotion. It was a primitive, barbaric war. The experience doesn't make me a very easy person to talk to, because I cannot be dispassionate about it."

Sitting in her apartment in the fashionable King's Road district of London, I have some difficulty in picturing this woman in mud trenches or aboard a battleship amid three thousand troops. She is a distinctive-looking woman of thirty-eight with short dark hair, dark intense eyes and a surprisingly low voice that belies the emotions she is expressing. "I'm the type who hates any physical discomfort," she says. "I'm even scared of insects. So for me the whole experience was even more striking."

It is not, however, striking for Linda to do unconventional things. The first was to leave an aristocratic home for the bohemian life-style of a London artist. "It was shocking," she laughs. "If I had done what I was supposed to do I'd be married to some bloody viscount. But to become an artist. . . . One simply didn't do such things."

Her talent brought her national recognition, a teaching post at the Royal Academy of Art and numerous shows and publications. Her artistic style often manifested itself in long studies of subject matter and voluminous drawings. Her work has consistently taken up enormous amounts of time, merging with and sometimes indistinguishable from her private life. In April 1982 she was in the midst of preparing a major exhibit of drawings on the workings of *The Times,* which was the culmination of a year's work recording life on Fleet Street and the fast-paced world of reporting. On 2 April, the day Argentina's forces invaded the small group of islands in the South Atlantic over eight thousand miles from Britain, the Falklands invasion was only another headline.

Though many of us had never even heard of the Falkland Islands, and the crisis seemed to descend out of nowhere, the dispute over their sovereignty had already lasted five hundred years.

In 1570 Britain claimed discovery of a group of islands that lie about five hundred miles from Cape Horn, but Spain protested that she had sighted them a hundred years before. In 1779, during the English-Spanish Wars, the British settlement on the Falklands was overcome by Spanish forces, and in negotiations that followed, Britain agreed to abandon the settlement.

In 1883 the uninhabited isles were resettled by the British, who would argue a hundred years later that sovereignty was determined by a long period of peaceful settlement. But Argentina, having gained all former Spanish holdings in the area with its independence, laid claim to the Falklands. During World War II, when England sought Argentina's alliance, the British discussed returning ownership of the islands to Argentina in return for a long-term lease for the settlers. After the war the talks were discontinued. Finally in the mid-1960s Argentina took the issue to the United Nations, where the General Assembly urged the two countries to come to some agreement over "this problem of colonialism." Seventeen years later, after fruitless negotiations, General Galtieri of Argentina ordered an invasion.

Prime Minister Margaret Thatcher responded swiftly. Stating that "agression must not be tolerated" she ordered the deployment of a one-hundred-ship naval task force to the South Atlantic. General Galtieri threatened a massive air attack on the fleet as soon as it entered the area.

The Imperial War Museum committee selected Linda Kitson a few weeks later as their choice for war artist, on the basis of her London *Times* exhibit. What they had seen in her work was an ability to draw people in motion, quickly and beautifully. There was only one problem: she was a woman.

"And women weren't allowed there. Not even the medical personnel could bring women on the grey funnel war ships. The director of the museum and I thought about this and decided to take our chances. I said if I could get as far as Ascension Island, which is halfway there and not considered 'war zone,' I would find some unofficial way to get to the Falklands. The director said, 'I didn't hear you say that,' and gave me the biggest wink on the side."

It is a British tradition to send an artist to war. The tradition holds even today, when film footage and photographs reach television and newspapers within hours. The British still consider the impressions of an artist of import: "Artists give us another dimension: the little personal details of being in battle, as well as the big picture."[1]

"It was a little bit crazy," says Linda. "I'm a pacifist, and the whole thing seemed wrong. And yet I felt this old-fashioned sense of honour that they had chosen me. It was all contradictions. But for an artist the chance was irresistible. I did get quite a lot of flak from my colleagues. Some of them said, 'As an artist, how can you go along with people who take up arms?' Well, I thought about it and decided it was exactly because I was an artist that I must go. My great belief is that everybody should see and experience everything. Otherwise how can they have grounds for deciding what they think about things? So it is central to my belief that one of the things an artist can do is to exercise their skill toward that end.

"The people who love me, my family and friends, were of course very worried about me going, and some tried to talk me out of it. But I didn't give myself time to really think about it. I did put together a detailed will before leaving," she adds. "I suppose deep down I wondered if I would come back.

"To be honest what terrified me the most was that I was setting a precedent — 'the first official woman war artist' — and I read somewhere in the press that I would be changing the face of contemporary art. I knew they could only afford to send one artist, so it all fell on my shoulders to capture as much as I could, and I knew absolutely nothing about war or the military."

Linda was advised to be discreet about her appointment in the weeks preceding her departure. That made it more difficult to figure out what she should pack and how to prepare for the assignment. "It was an unsettling time," she says.

The islands consist of approximately four thousand square miles of hills and bog. Linda was told only to prepare for both equatorial and arctic conditions for an unknown length of time. Winter was approaching, and a major concern would be keeping her work dry. She gathered folding stools, large rolls of plastic waterproof wraps and a fisherman's parasol that provided a circular cover about ten feet in diameter. Clothes were another problem. The voyage to the islands would take ten days. Having been given honorary officer's status, she would have to dine at the head table, "under three thousand critical pairs of eyes."[2] Her London wardrobe of old-fashioned opera shoes and punk jackets would eventually be discarded in favour of requisitioned army fatigues and thermal underwear.

On 12 May Linda boarded the requisitioned passenger liner, the *QE 2,* which was headed for the island of South Georgia carrying three thousand troops from various detachments including Welsh Guards, Scots Guards, and the Duke of Edinburgh's own Gurkha Rifles. (An enormous vessel, considered the fastest passenger boat afloat, the elegant liner would require a $10 million face lift after its war service.) On that same day the British fleet neared the Falklands and General Galtieri declared the entire South Atlantic a war zone.

Linda's presence aboard the ship was met with some disbelief and even hostility. She later wrote in *The Falklands War: A Visual Diary* that "any objection to my presence, however, was so frank and artless, I found it disarming. And by the time I had been around for a while, I was accepted. My attention, after all, was concentrated on them and *anyone's* attention is welcomed in bleak times."[3]

Learning military language and protocol was the first step. "I didn't know where to start — I didn't even know the rank structure. I had to have a little chart to identify what rank I was talking to. The most extraordinary thing was what was happening to the ship. Two decks of swimming pools were being converted to helicopter landing pads. The liner's shopping arcade was mounted with artillery guns for drills, and headquarters was set up in one of the boutiques. I asked a gunnery sergeant, "Aren't you from perfumes?' and he told me very primly, 'No Ma'am, we're in cosmetics.' It did have its lighter moments."

But Kitson's main concern was staying out of everyone's way as three thousand troops were being "whipped into shape." "I went to work immediately, drawing everything and anything that moved. Everything was new to me. It all looked fascinating. On deck men were lined up in rows on their stomachs lining up the sights of the anti-tank weapons. There was riflery practice. The noise was incredible. Helicopters and planes were taking off from the flight decks.

"I was given official access to every part of the ship, and the commander helped make sure I knew everything about the different activities. He and his men organized my time and took me to all areas."

There was a lot of socializing at night, and Kitson had a list of invitations tacked on her cabin door every day. But "it was also getting very tense. It grew worse as we approached the Falklands. I saw sides to men I hadn't seen before. They are very competitive. Some of the units were not used to the intense physical training — they were usually just ceremonial guards. There was an awful lot of slanging of other units, which I thought was quite hostile. I was told that it was just their way of psyching up. It was very male, from the most trivial comments and pranks to things that I thought were fairly unfunny. But it was necessary, they told me, and not to be worried about.

"On board ship we had little printed newspapers with jokes in them. Humour was at a premium. Things like a labour strike back at home struck the soldiers as terribly funny. It seemed so petty, the things people fight tooth and nail for at the civilian level, when you were on your way to war."

As the *QE 2* sailed farther south the weather became colder and the waters rough. The doctors issued seasickness pills, which Linda "thanked God for." As the temperature dropped rapidly she layered on regulation army clothes, but stood out in the crowds because of her wildly striped leg warmers and woollen scarves. By now she had made some close friends and told them about her plans to try to get all the way to the Falklands despite regulations against women's being sent there.

"After we'd been on board six days the soldiers were determined to keep me with them," she laughs. "There were two or three ploys thought up by different ranks. I have no idea what the plots were to entail, but it would be easy to hide among three thousand troops. I thought that it was a very nice tribute to art that they didn't want to lose their artist."

As the *QE 2* neared Ascension Island Prime Minister Thatcher ordered her naval commander to prepare for an invasion of the Falklands. Because the war was escalating so rapidly, the liner was ordered to bypass Ascension, where Kitson was to have got off. "We went steaming on past Ascension and stopped for refuelling at another little island," she remembers. "Officials came on board who could have ordered me off there, but it came through higher command that I could land with the men. Someone had decided to make the exception."

Once past Ascension the atmosphere on board sobered. Mail deliveries were infrequent, which noticeably lowered morale. "If I hadn't been there I wouldn't have believed it," she wrote. At this time all personnel became officially on active duty.

The battle in the South Atlantic began on 22 May when the British launched an amphibious landing about fifty miles west of the Falkland capital, Port Stanley. It was quickly followed by landing ships that carried Scorpion light tanks and anti-aircraft missiles. Seven weeks after the Argentinian invasion the British regained a hold on the islands.

On the *QE 2,* training intensified. In anticipation of air attacks that might fill the ship with smoke, the Gurkhas were trained blindfolded. They jumped into the freezing swimming pools dressed in combat gear. On a deck below there were training films, slides and lectures. In the most current reports the troops were informed that ten thousand Argentinian troops had been deployed to the Falklands.

Kitson later wrote, "Deck-bashing [an exercise in which fully kitted soldiers ran back and forth across the deck] started before seven-thirty every morning and continued throughout each and every day. The units booked deck space in rota and formed a human chain, pounding ceaselessly right around the boat deck and up and down the upper staircases.

"The men carried up to 120 pounds. Some carried each other upside down; some carried heavy weapons and equipment. Sweat poured off them. Once moving, they could not have stopped. To walk in the opposite direction would have been suicide. Strangely the soldiers appreciated the exercise and felt better after it. They suffered from the constraints of a crowded ship."[4]

On 28 May the *QE 2,* reached Grytviken, South Georgia, in sub-zero temperatures. There they were to transfer to the liner *Canberra.* Linda gathered together the drawings she had done, wrapped them in plastic and left them on board the *QE 2* with instructions for their delivery. She described the transfer in her book: "The helicopters flew back and forth to the ships all day and night, carrying supplies, baggage, troops, and myself. . . . The long-stay vessels at Grytviken shrank to toys when seen beside the hulk of the liners."[5]

"Once on board the *Canberra* I was told I had to participate in the fire and lifeboat drills. For the first time I really panicked, thinking I could never survive it," she says. "I was also worried that I was becoming a nuisance, because I noticed in fire drills, for instance, when your life depends on getting to the lifeboats quickly, the men would open the bloody door for me. And I thought, 'Well, this is funny, but it must not happen in battle or campaign, because then I would actually be detrimental.' They were very gallant, and old-fashioned chivalry prevailed at all ranks."

Linda says that she was perpetually scared as they neared the combat zones. "It was an awful pit in the stomach, a sort of great hole with nausea, very real to the point you thought you were going to be sick. And there was endless fatigue because by this point I was hardly sleeping at night, and it all just built up and built up." She asked some of the soldiers if they were scared too. "I think they mainly just wanted to get it over with," she says. "But when we talked about it afterwards, I found that they had all been feeling exactly the same way. We had made this extraordinary voyage — the weather was dreadful — so we all felt awful the whole time."

After three days on board the *Canberra,* on 2 June, they reached Falkland Sound. It was a foggy, bleak day and only the outlines of the battleships *Fearless, Baltic* and *Nordic* could be made out through the mist. The troops on the *Canberra* waited fully kitted for their orders.

Within hours they disembarked onto a landing craft as Linda watched from above. She stayed on board another day, "lonely and frightened," while the *Canberra* took on Argentinian POWs and British wounded. The following afternoon she was taken by lifeboat to the battleship *Fearless* for a meal, and then by motor dinghy at nightfall to the shore where the first amphibious troops had landed ten days before.

Those first three thousand troops and now Kitson landed on Pat Shore's sheep station, a rough piece of land bordered by hills and a rugged shoreline. Little white buildings stood out in the moonlight. Linda was boarded with the family, along with men of the 16 Field Ambulance, RAMC.

In the morning she was taken around the sheep station, which now was converted to "40 Commando Valley," a base camp made up of "all styles of dug-outs and huts riddled with shrapnel holes. The hills were slippery with mud and I immediately fell down them. My first impressions of the islands fitted exactly a description from a poem written in the eighteenth century. In it they were described as a terrible piece of land, just blasted barren rocks. The landscape looked Scottish, complete with thick bogs and wretched weather. The poet had written that one day these rocks would cost a terrible price. I thought, 'How did he know?' "

Linda was assigned a marine lieutenant to look after her, whom she soon began calling "Hawkeye" because "he was always on the alert for any danger." She was told what alerts were the most serious and given a tin hat to wear as she was escorted through the base.

"The whole scene looked very much as I imagine World War I must have looked like," she says. "Mud and trenches. It was so primitive. There were sophisticated air-to-ground missiles, and ship missiles and things, but once on land all they had were light field guns. Nothing else would move. There was simply no way to move more than a few kilometres before getting bogged down." Everything at 40 Commando was delivered by air. Only bulldozers were still moving through the mud on the base.

Troops were preparing to move towards areas of intense fighting. "The thing I found catastrophic was the strength needed to lift the 130 pounds they carried on their webbing, their shoulder straps, with bags, helmet and rifle. That's a lot of weight, and to traverse that terrain they had to be pretty crack outfits. Some were carrying mortar platoons themselves to be able to continue the battle. Every helicopter was doing — God knows — treble time just to keep things moving."

If the fighting on the ground seemed primitive, in the air above it was technologically sophisticated, supersonic warfare. The Argentinian air force attacked with French-built Super-Entendard and Mirage fighter-bombers. The British used Harriers and computerized Rapier surface-to-air missiles. Linda was not in direct combat but worked amid constant air raids and alerts. She soon began to act like the Falklanders, ignoring all but the most serious alerts.

"Every day I was taken to some new site or mission," she says. "God, those bloody islands made just walking difficult. The wind and hail would come up so that if you didn't have your face covered up it would look as though it had been slashed by razors. The sleet cut right into it. Everyone very quickly learned to cover up. The sight of me making my way across base camp, carrying a portfolio which would act like a sail against me, going knee deep in mud every time, caused every kind of reaction. Even if nothing happened other than that I, as I did hundreds of times a day, fell flat on my bottom and slid down in the mud, it would cause a lot of hilarity. I would be picked up by about twenty people, dusted off and sent on my way. If there were people free I could hardly walk two steps without someone coming to help me carry something."

After five days at 40 Commando Linda said goodbye to the Shore family and the men stationed at the base, and was taken by Seaking helicopter to Goose Green, where one of the bloodiest battles of the war had taken place.

"It was there that I had to decide which aspects of the war I wanted to cover," she says. The recently vacated Argentinian trenches now housed the Gurkhas, whose mission was to clear the area of bombs and napalm stores that were scattered through the village. Houses were converted to billeting stations, their yards covered in anti-aircraft guns and missiles, their washing lines draped with fatigues.

"My brief was to record the sights that might be considered as common experiences, so I chose not to draw the gory aspects," she says. "I walked by them every day. I saw the horrors — the helmet with a head in it, the corpses. I passed by them, but always en route to another assignment. My days were very tightly scheduled. I was always being taken to some important operation across a mine field or up a mountain side. Instead of the other horrors, I recorded the men in action."

Kitson had decided to capture the human story of the young soldiers, whom she was becoming fond of. She described their depressions, fears and boredom as similar to her own. "It became the norm," she says. "There wasn't anybody around that you could look at that you didn't look into their face and think, 'Oh, you feel the same way I do.' In a way, had there been smiling, happy faces around I probably would have felt worse. But since everybody felt that way, it minimized it.

"I learned that the very worst times were when you were left alone and had time to realize how bad you really were feeling."

But she says she was happy to think that her presence there helped raise spirits. "When things were not actually hectic or dangerous I think there was a certain amount of enjoyment of me and the rather bizarre way I looked. They treated me as everything a female can be — mother, sister, friend — and I liked that. There was nothing sexy about it, really, under five layers of clothing, dirty, wet and all that. But every age and every station told me that they really like having someone to look after.

"When I went into some of the awful accommodations they had there in their dug-outs or freezing-cold barracks, I could see the effect I had on them. They were there, having had no mail, and not enough food, and deprivations like no cigarettes. First mail, then cigarettes, then food were missed, in that order. Mars Bars were on the top of their food list. Morale got very low because communications were terrible, and people got stranded in really dreadful places where you just had to hang on to yourself to keep going for the next half hour. No one knew how long they would be there. If I came around everyone would sort of straighten themselves up and stop swearing at each other, or do it for fun. It was because I'd come in. It was nothing personal about me, just the presence of a woman. It really did change the atmosphere. Afterwards, those who were not too shy told me it makes men feel more human if there is a woman around."

Kitson worked long hours in the freezing weather, drawing everything she could at Goose Green and on missions in nearby areas. From a former Argentinian trench she drew the prisoner-of-war camp, a sorry-looking collection of wooden barracks across a barren wind-blown field. She drew the Gurkhas in their dangerous missions of clearing the area of explosives, the highlight of which was the setting off of a huge store of napalm. She drew with "fighters screaming overhead" and "in between alerts." Her drawings of the Goose Green airfield reveal the remnants of fourteen Argentinian aircraft. She drew medics fast at work tending war casualties and victims of the severe weather conditions.

"I had my own battle going on," she smiles. "A different one. I was going to say a lesser one, but I must not say that as an artist. Still, in light of the fighting going on it did seem to me to be lesser. I didn't really have time to think about what was going on around us. I don't think I wondered what I would do if someone were killed beside me, if I would actually take up their rifle to defend myself. There was no time to consider it. Do you realize what it is like when you are sitting in a trench and your fingers have gone through the agony barrier, and you've lost your feet, they've gone, and you are going to have to have help being moved from the position you are in — to try to draw? In many cases I would just do very slight drawings that I planned to finish later. There wasn't time to do more.

"The difficulties of being a woman were physical disadvantages. Having a period or even having to pee became a very complicated procedure — especially with five layers of clothing on. But all those things were very minor considerations compared to the pit of fear I shared with all the men."

The troops at Goose Green were in final preparations for the retaking of Port Stanley. Linda was working under constant red and yellow alerts.

"The only thing I kept worrying about was not to be a burden. The men were facing the ultimate danger of death, or the ultimate fear which I noticed for most men was the fear of being maimed forever. They really were more scared by the thought of a major physical setback for life than of dying. To have been party to any such disaster would be something I could not have forgiven myself for. So I went around an apology for living. It's one of the arguments people have about women in battlezones.

"In the airfields they would come out of cover, which they are absolutely not allowed to do, to give me a tin hat. Which was dreadful. I always made sure I got my own. But they would risk that for me."

Logistics of eating and sleeping were difficult. Linda survived mostly on hard-tack biscuits and, when possible, coffee. "Emotions were raw and nerves shot." Then came the death of Major Mike Forge, who had taken Linda under his wing the moment she boarded the *Canberra*. "He was blasted out of the sky in a helicopter. It devastated me. He had been responsible for instructing me on everything I knew about survival, both on ship and on land. It is still a shock. It only took the death of one person, someone I cared about, for all the suffering around to hit me. The terrain was so flat that I had seen a lot of death in the distance, but this was someone I knew. My reaction was a spontaneous weep. The men wept too.

"Death so quickly became the norm that I got hardened to it," she reflects. "It was that first one that was the hardest."

"At night there was a lot of drinking. On one occasion, having spent the day freezing in a mine field, I got back to camp and took a couple of hard drinks and keeled over. I ended up in the field hospital, where I spent the night."

Kitson says that she became more and more respectful of the soldiers she was getting to know. "They were taking all this, and still their training was seeing them through it. I came across any number of soldiers who were horrified at the idea of killing. They are described as 'trained to kill,' but for a lot of them it was a very difficult thing.

"I don't think it was patriotism I was feeling. I was just filled with sympathy and admiration and rather human reactions to the people I was seeing coming down off mountains where they had literally been fighting with bayonets, hand-to-hand. And when I asked the boys before leaving how they were feeling, believe me, it wasn't anything glorious like patriotism. It was, 'Let's get this thing over with,' and much more basic things. When you are in those conditions it's much more, 'Is my arm all right? Is it still there?' You aren't thinking anything grand. The grand thoughts come later. Out there it was just half hour to half hour and how to keep going."

The days 8 and 9 June brought tragedy to the British with the attacks on the ships *Sir Tristram, Sir Galahad,* HMS *Plymouth* and HMS *Hercules.* Linda remembers: "There were men pouring off the *Sir Galahad,* and oil leaking from it on fire, coming towards them. Nobody gave a damn what unit anyone was from. They all tried to save the others. The helicopters flew in and out of that smoke with their propeller blades so close that if they'd touched the ship they'd have gone down. The wind from their propellers would propel the burning oil away from the men. When they landed it was a question of getting blankets, getting those people who couldn't walk onto land safely. People were on fire, screaming. And I saw men go back into positions where they could be killed instantly, reach out to help someone when really you'd think they'd be out to save their own hides.

"That was the great shock for someone coming from civilian life, which is so much 'Up you,' look after yourself first, nothing but self-seeking and climbing up whatever ladder. To see, when things were really terrible, people doing just the opposite, endangering themselves again and again for another man. It just obviously came so naturally to them. You help the bugger. You bring him in."

Thirty-nine Welsh Guardsmen were burnt to death on the *Sir Galahad* that day, and another nineteen were seriously wounded. The ship burned for another week before it was sunk as a war grave in a ceremony that Kitson described as "almost more than I could stand."[6] In Britain, as the public paid tribute to the dead, editorials examined the controversy over leaving the ships unprotected in the cove, a result of the generally hasty planning that had been a result of unexpected war. Prime Minister Thatcher hounded her military commanders to move quickly to end the war before the full fury of arctic winter further bogged down the British troops.

From the headquarters at Fitzroy preparations were made for the final British attack, aimed at Port Stanley, fifteen miles away. It was to be a frontal attack against the remaining Argentinian positions on Sapper Hill overlooking the largest settlement on the islands. The fighting was intense. Argentinian troops were pounded from their positions by Rapier fire and heavy artillery, an assault of infantry and paratroopers and sustained air support.

The Argentinians surrendered on 14 June. That evening cease-fire was announced and the papers were signed in the headquarters.

That morning Linda left Goose Green for Fitzroy by plane. She wrote, "The news had come through of a white flag flying from Port Stanley harbour area. We tried to make ourselves believe it was another trick, for we knew that the previous night's fighting had been very intense."[7] But upon arrival at Fitzroy she heard the news confirmed.

Linda's work continued much as it had before as she battled

the elements to draw the clean-up missions that followed. Fitzroy's conditions were worse than those she had left at Goose Green. The settlement was surrounded by low-lying hills where the fighting had occurred, and now ten thousand men descended into the base suffering from frost-bite, trench foot and hypothermia.

The problems of the aftermath were recorded in her drawings. Troops remained in dug-outs in the hills for the next few months, warmed only by small heaters. Finding shelter for troops awaiting passage home had become a logistical nightmare. In an interview with the London *Times* later, Linda said one of the worst sights she witnessed during her time in the Falklands was that of six hundred soaking wet Scottish troops taking shelter in a tick-infested sheep shelter that was colder inside than out. Shaking with cold and exhaustion, "they just hung on from one moment to the next."[8] A large stone building was requisitioned for luckier units, and it became the place for parties at night.

Her drawings now portrayed the very real difficulties of clearing the beaches of mines "that had been swept about by the wind all over the place. The wrecked bulldozer, blown up by a mine the previous day, was actually blown into the sea. . . ."[9]

On 12 July Linda moved to Port Stanley and joined a battleground tour organized for the press and the high command. In sub-zero temperatures she was happy to accept a lift up Sapper Hill in a captured Argentinian Mercedes.

Linda spent her free time talking to soldiers who did not know when they would be sent home and listening to their accounts of the battles in the hills. "It was an uncivilized, primitive war," she says. "We were dealing with an enemy army that was totally unethical. Their treatment of their own men was barbaric.

"And you might think, 'Oh, the British are so damned organized.' But I don't think it is crazy to try to keep law and order in the battlefield. It's how you keep your men together. It is, literally. You could say we didn't win it, but they lost it, because they didn't have law and order and their command was terrible. They intimidated their own men, abandoned them on the hillsides. The morale of their soldiers was so poor that I think it was a major contributing factor to their surrender on Sapper Hill. They just couldn't take the fighting anymore.

"I was told that up on the mountains, if the Argentinians showed a white flag to surrender, they would open fire as our men came to take prisoners. Thereafter the British command gave a ruling stating not to take notice if they surrendered. I had a long talk with a young man in the Scots Guards who could not get over the fact that his training taught him to put down his arms and advance, and yet his sergeant told him to continue firing. He described how he saw Argentinians with white flags beheaded by the spray of bullets. He's out of the army now. He couldn't bear that vision.

"I now believe that you must have rules and regulations in war so that you know what to do in predicaments like that."

Like everything else during those three months, Linda's orders to return home came suddenly and unexpectedly in the middle of an approaching storm. "I couldn't believe we were really going to fly," she remembers. "The weather was so bad that the helicopter flight to the ship was terrifying." She was depressed to leave "when so many of my friends remained behind." The twelve-day voyage home was made in the company of the Welsh Guards, who had been so badly hit on the *Sir Galahad.* They were returning home without fifty-eight of their unit. It was a sober journey.

As the ship approached Land's End, England, rumour spread on board that Prince Charles would be there to welcome the troops home. It was met with mixed reaction: "Many of them just wanted to slink away unnoticed, in sympathy for the widows. It was a very close unit, and so many of them were gone. There was a lot of controversy about getting special attention. But when we landed everyone seemed to convert at the last minute. We had all felt as though we were never going to get back home, after all those weeks sitting around Port Stanley, seeing all our ships being used to transport the Argentinians. ('Typically British,' she says. 'We win something and then pay forever.') But finally we were home. We looked down at Land's End and saw England, and it was very emotional. And then it did seem right, after everything, even to the most badly hit, for

Prince Charles to be there, because he knows all about combat. He flies and sails and does all those things, and he is heir to the throne. It made the boys feel, after such neglect at Port Stanley, that they were recognized.

"I can't remember a word he said to me — something about the safety of my drawings, I think. But from that moment on I was deluged by the media."

Linda had been away for three months and brought back over four hundred drawings. She was exhausted, had lost a great deal of weight and, as a woman used to grasping things quickly, had been thrown off by the emotionally confusing experience of war. The press descended on her in droves. "There was tremendous curiosity about how a woman got on," she says. "So I got double the press. They seemed most interested in what attitude the men had to me, and also in the fact that I could have survived the experience. They were interested to find that I had brought back such a large amount of work, clearly proving that women can undergo this. People would think I was some kind of Boadicea or something, Journalists would come and say, 'Ah, we thought you'd be a great big hefty thing.' At that time I was terribly thin. They couldn't get over the fact that I wasn't huge."

"Did the press bother me? I was so overwhelmed by feelings of tribute, no matter what I thought about fighting and war, by proximity to those men, that I did tend to go on and on in the interviews saying that those men had done something tremendous, heroic. Since the military aren't allowed to give interviews I thought I'd say it for them. It had nothing to do with the issue of whether or not I agreed with the fighting of that war. I wonder if this is a female thing, that I didn't have the ability to remain detached. I am very emotional. In retrospect I think I was too fervent about how terrific the boys had been and didn't say enough about the horror. Everything I said was used politically as a tremendous feather in the cap for Britain, which wasn't what I meant."

Linda spent the following twelve months working on the Falklands drawings, adding more detail to some, and preparing for the subsequent exhibit and publication of *The Falklands*

War: A Visual Diary. Her work took up all her time and kept her thoughts on those three months. As she began to examine her drawings later, she decided that perhaps she had made a mistake in not portraying some of the horrors she had witnessed.

"I think I made a mistake," she says. "It was a tremendous loss and a grave miscalculation that I didn't do those kind of drawings. Maybe it was an instinct to defend people from the very worst, to avert their eyes. I don't know. I could have gotten the permission if I'd known then and asked through the right channels. I was kept quite separate from the journalists, and there was a degree of trust towards me which would have enabled me to draw anything I wanted to. Later I met some of the Welsh Guards who were some of the worst victims of *Sir Galahad*. They had been a terrible sight in the hospital — defaced, their skin falling off, dismembered. I had not drawn them because I felt it was very important not to hurt them any more, not to cause them offence by drawing them when they were like that. But when I spoke to them later they told me they wouldn't have been in a state to know what I was doing, and even if they had been conscious they would have been proud to be portrayed like that.

"I wish I had done it — to counter all the films which are so successful in portraying war as a grand venture, thrilling, exciting. We know from the success of these films that there is a tremendous amount about war which people really do love. I don't think they would love it if they saw it. . . . I should have used shock tactics so they would know."

But the directors of the Imperial War Museum and the Fleet Air Arm Museum had no misgivings. They felt that Linda had managed to capture what it felt like to be a soldier in the Falklands in a way a photographer could not have done. Because the battles fought over the islands had almost all occurred at night and in the mountains, her drawings do not include scenes of the primitive battles that made the Falklands such an unusual war for its age. Instead, the images of Seakings and burning ships seen from afar, pieces of machinery half buried in the mud, and soldiers cold, wet and miserable in their dug-outs are what Linda has portrayed. Despite the absence of "shock tactics" her drawings do not glorify war. They personalize its many moments of boredom, discomfort, fear and despair.

In the years since the brief Falklands War, Linda has worked intensely and taken on another project involving war — this time drawings for a film about Cambodia. Her confusion over her personal decisions in the Falklands, her emotional reactions to her experience there versus her pacifist beliefs, have not abated. "Now I think I am glad I went," she says, "but if I were to do it again I would do it better. I would draw and talk more about the ultimate damage to people. I saw people give, people who knew that the whole thing shouldn't have been — at a young age taking on a very brave attitude that overrode the knowledge in their minds that it was a political gambit. I saw heroism and waste.

"It has changed me. Now my closest friends are some of the people I knew there, maybe because it was difficult for my old friends to understand me. Until you have experienced something yourself, it is impossible to understand, I shouldn't have expected people to know what I was going through. I tried very hard to limit discussion about it to those who were there. But sometimes I would be staying with someone and break down in the night crying, or have a sudden breakdown in public, which happened particularly when I was trying very hard for it not to happen. It is an involuntary thing and depends totally on the state of health you are in. There are lots of manifestations of distress. Sometimes people were confused because it was outside their domain. They had no sympathy for it, sometimes thinking I was trying to bring attention to the fact that I had some experience they didn't. In general people weren't very forgiving of my behaviour when I first came back and was trying to adjust."

Linda has discovered that there is no time limit on post-traumatic stress. She says, "There is nothing as big, or as important, to replace that experience in my mind." Hearing how her companions in the Falklands have fared since coming home also keeps her thinking about the war.

"But one thing is comforting to me," she says, smiling. "I've gone on about all the things I didn't do, or didn't say — but I think I did something. After the show and after my book was published I got about fifty letters a day. So many of them said, 'What you did meant more to us than anything else.' I got letters from the bereaved saying, 'There is nothing else for us to hold on to,' and from the boys who left the army afterwards, saying, 'The only thing we want out of the whole damn thing would be a drawing.' So it did reach people. It was my promise to the boys, when we were in the Falklands together, that someday there would be a show, and they could finally see some record of their actions."

ANOTHER REALITY

El Salvador, 1975-

We had some problems with chauvinism in the beginning. The men didn't want us to join, or they wanted us to stay in subservient roles. But soon they realized the importance of having as many people fighting as possible, and they changed a bit. I think it actually helped make male-female relationships more equal.

Ileana,
Urban guerrilla, El Salvador

I used to carry a huge bag with all this kids' stuff in it — talcum powder, diapers, baby bottles. I hid my gun and pamphlets underneath. . . . The soldiers never thought of checking in the baby's diapers.

Maria,
Guerrilla, El Salvador

The two women who speak in this chapter were interviewed in Toronto, Canada, where there is a growing community of political refugees from Latin America. They have asked me to change their names to protect themselves and members of their families who are still in El Salvador. I will call them Ileana and Maria. At the time of this writing both women are twenty-five years old, both are mothers and both have spent the past decade fighting for the revolutionary forces in El Salvador.

The similarity does not end there. Both women have been arrested, tortured and imprisoned, and both were released because of internal and international pressure on their behalf from organizations including the Catholic church in El Salvador and Amnesty International. Each woman has lost her husband: one killed by a death squad, the other by official government forces. Finally, both Ileana and Maria were sent to Canada to recuperate.

Ileana has now returned to El Salvador to continue fighting. She was in Toronto for eighteen months, and during that time she worked for the Farabundo Marti Front for National Liberation (FMLN), seeking audiences with politicians and concerned citizens. Now she has left her infant daughter in the care of a Canadian family and returned home. Maria has arrived only recently and is still recovering from the scars of two and a half years at Ilopango Prison. She was unable to walk for six months after her release.

The women differ both physically and in character. Ileana is the more political. She gave up the religious beliefs that first got her involved in protest groups when they ceased to help her understand the tragedy of her country. Now she believes in socialist revolution. She is a small woman with large dark eyes and a soft voice. She wears no make-up, and nondescript clothes. It was difficult to get her to talk about herself. She still cannot talk about her husband's death because it is too painful. She softens considerably when she speaks of her young daughter, and when she finally agreed to tell me her own story she became very emotional.

Maria speaks no English and is still adjusting to being in Canada. She is still reeling from the blow of her husband's death and is deeply concerned about the welfare of her two small children, who have remained in El Salvador. Maria is more feminine in appearance than Ileana: she wears make-up and colourful clothes. She looks small and fragile; she is very thin and looks impossibly young to be a mother or to have such a long and bitter personal history. Still a devout Catholic, Maria believes that revolution in El Salvador has little to do with Marxism. "There are Marxists and socialists fighting," she concedes, "but there are also many of us who believe in democracy. We have united because revolution is the only way to end the tyranny."

It is important to keep in mind that these conversations are about things that are going on today and that these woman do not have the hindsight that is available to the other women interviewed for this book. They are still experiencing or preparing to experience again the situation they describe. The war is far from over, and they cannot be anything but partisan about it or their roles in it.

The war in El Salvador is a passionate subject, going back a hundred years to the 1880s, when the government decreed laws that recognized only private property, thereby effectively destroying the traditional communal landownership of the peasants. Illiterate peasants were never truly informed about the new laws, and the rich bought their land out from under them and turned it into coffee plantations, which produced wealth for the oligarchy but left the peasants with only seasonal labour at pitiful wages. Worse, coffee growing meant less land for growing food. The rich imported most of their food. The poor starved. Today the rich largely control the government and an economy that is described by experts as the most inegalitarian in that part of the world.

It is passionate, too, because it is a struggle that we cannot fully understand; some believe it is a Marxist-dominated revolution instigated abroad and a threat to Western security, while others believe it is the natural uprising of an oppressed people against a series of corrupt and brutal governments. The facts remain that El Salvador today is a country in which seventy-five percent of the population have no land, no education and not enough to eat. Obscene acts of violence and murder are daily perpetrated by the ruling government against nuns and priests, pregnant women and even children.

The women's actions, as always, speak for themselves. The women are hard because they have to be, because they have survived combat, rape, torture, imprisonment, fear and loss of loved ones. They are hard because there is no time or place to be soft, no time to indulge in the day-to-day life that people in peacetime perceive as reality. They do discuss issues like sexual discrimination, and see already that the fighting has given Latin American women new roles. But these are not issues that concern them greatly at the moment. They are interested in talking to me only because they want their stories told. They want people to know at least their side of the reality that is El Salvador at war.

Ten years ago Ileana was a fifteen-year-old high school student in San Salvador, a deeply religious Catholic, reared in a middle-class home, who liked rock and roll, boys and nice clothes. Today she is a committed revolutionary, living underground, rotating between the city and the countryside where she carries out organizational work and participates in armed actions against the government.

"I believe now that armed confrontation is the only way for El Salvador, even if it is the most painful way, because all the political expressions of the people have been suppressed, all the peaceful means of protest have been attacked and even our archbishop was killed for his views. What other way is there to change things? The only way left is to pick up our guns and fight for a better life."

After eighteen months of safety and rest in Canada, her choice to return was in many ways more difficult than her decision to become a revolutionary in the first place, because now she knows all too well what being there means. "It frightens me," she said before leaving, "because here I've grown used to being safe, and I've lost many of the instinctive defence habits I once had in El Salvador, the things that I used to do automatically to keep safe. Here you can walk down the street day or night, sit in a restaurant, talk about politics, and your life is not endangered. I feel that I need more preparation, that I'm no longer ready to face the confrontation. I have to find the strength to leave my daughter, knowing I might not return, and then both her parents will be dead."

She looks at me with wide eyes but speaks with the hardened resolve of a veteran revolutionary who has robbed banks, kidnapped men and killed. She has done these things because "I believe that only revolution can bring about the changes we need in our country."

The guerrilla war in El Salvador has veteran Vietnam correspondents shivering with *déjà vu* in the tropical jungle. American involvement continues to escalate in the region, and there is equal evidence of aid and military advisers from the Socialist bloc. Families are divided; sons and daughters fight each other. The television images are hardly distinguishable from the scenes the world watched not so long ago from Southeast Asia. But comparisons with Vietnam serve only to remind us that protracted civil wars become part of wider geopolitical cold war — and that millions of people die.

Ileana's father was a politician and businessman throughout his active life. His beliefs were to greatly influence the fate of his family and the choices made by his children. "He was involved in the struggle in his time," she says. "As a politician he saw too clearly the poverty and desperation of most of the people of El Salvador. He spoke up about it and was imprisoned and "disappeared" several times. Finally he was exiled from the country. El Salvadorans have been opposing the regime for decades. In 1932 the peasants rebelled against their low wages and loss of employment. Thirty thousand were massacred by the military and landowners. Because of my father's activities in the 1950s, we went to live in Guatemala for a while. Ever since I was a child we have lived in fear of what was going to happen to us. After we returned from Guatemala my father decided not to get involved in politics anymore. He could not take imprisonment or torture at his age. When we kids became active he told us to be careful. He is scared for us, but he supports us."

Ileana's mother was never really involved: "She really didn't understand anything, even after all those years with my father. Now they are separated, but even when she was with him she didn't understand. Still, she is very goodhearted and always helps us when we need her."

There had always been revolutionary currents, predominantly stemming from the middle class. The seeds of the present state of civil war were sown in 1970 when university students and professors began to protest against the excesses and corruption of the government, including nepotism, electoral frauds, pay-offs by the rich for legislation of benefit to them, and suppression by threats, torture and murder of opposition politicians and clergy.

In a country where the vast majority of the population is Catholic, priests and nuns working among the lower class also became opponents of the regime. They began to organize Christian-based rural communities that were in effect communes where peasants could collectively manage such needs as medical services and education. It was through church groups that Ileana and Maria became involved in their war.

Ileana remembers: "I was religious then, at fifteen. I belonged to a Church group and did social work in the community. It was at that time that many of the priests began to preach that our rulers were not good. The Church became more and more involved in the problems of the people — not spiritually, but in day-to-day life — and then they began to talk about it."

It was 1975 when Maria, also fifteen, joined a Church group in the town of Aguilares. "My father was from the lower middle class," she says. "He works as the overseer of a manor, and so from childhood I lived on manors and saw how the peasants lived, the kind of work they were subjected to and the problems they had just to feed their large families. I started studying in the nearby town of Aguilares, and that is when I first came into contact with Father Rutilio Grande."

Grande, a Salvadoran Jesuit, had arrived in Aguilares in 1972. In the rich, sugar-growing regions outside the town the peasants were unable to get more than three or four months' labour each year, back-breaking work that helped provide the average annual income per family of seven hundred dollars. The rampant malnutrition of their children affected him deeply. He and three other priests began organizing rural communities and preached that "they must not live in conditions of such tremendous inequality that the very Fatherhood of God is denied."[1] During the next four years Father Rutilio led peasants in several strikes and sit-ins at the haciendas in the vicinity.

Maria remembers the effect Grande had on the people: "He told us that in the Bible it says that people must not be exploited, people must not be oppressed. A hacienda in the area was an example of the way people lived. The peasants who worked there during harvest season were starving to death, and the rich widow who owned the estate would not allow most of it to be cultivated because she wanted it to be a memorial to her dead husband. The peasants had no running water, no electricity in their one-room huts. They had no education for their children, no medical care. Father Rutilio and three co-workers visited the widow and suggested an agreement for share-cropping in which the peasants would pay her for the use of her land. She refused, and so we organized a take-over of the hacienda. About two hundred peasants simply squatted on her land. We put up armed sentinels at different points, and we were prepared to confront the authorities.'' Maria was armed. Asked about weapons training, she said she had already been taught to shoot by her father, for recreation and hunting on the hacienda. Her parents believed her still a student who dutifully came home on weekends. But she had quit school and devoted all her time to the cause.

It was May 1975. The peasants remained on the estate for three months. They had begun to cultivate the land when the area was surrounded and attacked by the Security Forces in the middle of the night. ''One of our sentinels had fallen asleep on guard,'' says Maria. ''The troops were already inside the estate when our second sentry gave us the warning. We had prepared for this and had our escape routes planned. Each of us who was armed was responsible for leading out a small group of peasants — about fourteen or fifteen people. Others were assigned to cover our retreat. About fifteen minutes after we heard the warning, helicopters came and began dropping barrels of flaming gasoline on us. Our plastic, plywood shacks caught fire immediately. A lot of people were badly burned. We walked . . . ran, through the night. We had to cross a river. One of the women was in labour. She gave birth during the night.'' Maria led her group to a prearranged location where they met up with the others the next day. The peasants were told to disperse and remain silent about their involvement in the take-over.

Asked if they weren't expecting that kind of reaction to the illegal take-over of the hacienda, Maria said, ''We were of course expecting the authorities to come and make us leave. What we didn't expect was that the army would attack us in the middle of the night with helicopters and flaming gasoline. I was very scared after that backlash, but my older brother, who was also involved, said to me, 'Sure you can stay home and Daddy will pay for your studies. You'll have something to eat. But what about the other people?' ''

Father Rutilio Grande continued his work among the peasants. He organized literacy classes in which Maria and other young students helped teach the children. But the local landowners had by now had enough of the priests in Aguilares who, they said, "were instigating class warfare," and members of the Christian movement realized that there would be further violence.

"We were given training, told what to do if we were captured and questioned, how to respond, to make up a story quickly. They taught us how to deal with many problems. Throughout the rest of 1975 we led demonstrations in Aguilares. We went to the high schools and got more students involved."

In the capital, San Salvador, dissent was brewing in schools and universities. In July 1975 the army attacked a student demonstration protesting the government expenditure of three million dollars to host a Miss Universe Pageant. The army blocked off the streets to those trying to escape and opened fire on the crowds. Twelve students were killed, eighty wounded and twenty-four "disappeared." Ileana: "It was a peaceful demonstration, and they attacked us with tanks! People were screaming, in complete panic. It was terrifying. We joined together with other protest groups and decided on a joint action. Several days later we took over one of the cathedrals of San Salvador. We did it to denounce the army and what it had done at the demonstration. We asked the government to state who was responsible for the killing, and that the chief of police and the army be forced to resign.

"The one thing we did succeed in was getting attention, letting people know the truth. Because all the newspapers are controlled by the rich people, when things like that happened they would write a lot of lies. For example, if the army killed one hundred people, they would write that three people died in a cross-fire between guerrillas and the army.

"This joint action was important because it brought several groups together — peasants, students, teachers, labour unions and people from the ghettos. We discussed a revolution which would change the whole basis of society. We had learned that it was impossible to achieve reforms in the existing government. They would never change. By 1975 we were thinking of armed revolution. Several political groups formed armed sections and began taking armed actions against the government. They began in the cities, and then more and more grew in the countryside. I left the Church group. It was good, but I didn't think it was doing enough. That is when I joined the Revolutionary Popular Front."

By the end of that year there were many groups of opposition to the government, several with their own guerrilla armies. Eventually the left wing and liberals united and formed the Democratic Revolutionary Front (FDR) whose military arm was the Farabundo Marti Front for National Liberation (FMLN), named after the most famous communist leader of the 1932 peasant rebellion.

By 1976 guerrillas were conducting kidnappings and robberies to finance their weapons expenditures. Ileana: "We organized several kidnappings of rich people in San Salvador. I know that in Canada, when you hear about kidnappings, it sounds very bad, but for us it is not so bad. These are very rich people who got rich by taking money from peasants for the last one hundred years. We felt we had the right to ask them for the money they had ripped off. Some of the kidnappings were to get our friends — political prisoners — released."

Ileana quit university to work full time for the front. "I was responsible for working with and developing the peasant organization. I had to move to the countryside, spend time with the peasants and gain their confidence. Then I would explain to them the necessity of becoming organized. I practically lived with them, only going to San Salvador periodically to get my things. I was living underground. My family knew I was involved, but they didn't know the details. I didn't want to endanger them."

Ileana's family were mostly involved or at least sympathetic, but many middle-class families have been torn apart by the war. Because the army has remained the most direct route to advancement socially and economically, many middle-class boys volunteered for it. Many families have children who are on different sides of the war. Ileana: "It has been the most difficult on the middle class. For the poor, it has been an easy choice — they know what the fighting is all about. But in the middle class the mothers are mostly housewives: they don't go out to work, they don't see what is going on. They read the newspapers and watch TV, and hear from the media that the guerrillas are wrong. When their children become involved it is very frightening for them."

Maria had not told her parents that she had quit school, nor did they have any idea that she was involved with Father Rutilio in Aguilares. Since she lived with her sister in Aguilares while at school, she let them believe she was still studying and reported home dutifully on weekends. Her older brothers and sisters all belonged to the front, and they had a tacit conspiracy to keep silent at home. In 1976, at sixteen years of age, Maria married one of the *compañeros* ("comrades-in-arms") with whom she had participated in several actions. She says she married so young because of the situation in her family home. "I had not been able to discuss anything with my parents. I couldn't tell them what I was doing, thinking or feeling. I was completely absorbed in the revolution and often endangered, and yet I went home on weekends and pretended to be a meek, obedient little school girl. Getting married was my freedom from this double life." She adds, "My husband was a few years older than me, and it was an intense relationship: we believed in the same things and took the same risks."

Shortly thereafter Maria had her first baby, a daughter. Asked if becoming a mother had led her to consider leaving the movement, to avoid the risks, she said no. "Having a child reinforced my commitment. As a mother I felt even more strongly about helping to create a new society for our children to grow up in." But only fifteen days after the baby was born, she and her husband were arrested in Aguilares. "We were caught spraying slogans on a wall. I was not mistreated, just interrogated. They asked me why I was doing this, who my family was. I told them my uncle was a colonel in the Security Forces and hoped that would carry some weight. After two weeks they released us.

"After we were released from jail we stayed in Aguilares for two more months. Then Father Rutilio told us we should relocate and begin living underground. The police had our number, they were watching us, so we moved to another district and went 'underground.'"

Ileana was also married that year, to a man she had met in the cathedral take-over in August 1975. He had subsequently joined the guerrillas. "For the first four months we lived separately. Then the organization authorized us to live together, but we were told to be careful because it would be dangerous for us to be identified together. That is because he was a known guerrilla, while I was still working in the open. I was only home one or two days a week. The rest of the time I was in the countryside. It was a difficult period. I felt as though I didn't have any home. I was always travelling from one place to another, sleeping here and there, and worrying about my husband while we were apart."

In a political climate where ten thousand people became *desaparecidos,* or "disappeared," each year, working underground in the city created enormous stress. Maria says: "It was very tense. We constantly had to change our names and identities. We were always on the move. Sometimes I would forget what name I had used with different people. I would run into someone and not know how to respond, desperately trying to remember what I had told them about myself. But the nature of our work gave me energy. I found it so rewarding. I got used to the pressure."

Maria's work was still with Christian groups. "We had contacts in different parishes with different priests and we would join them in their discussion groups. We worked in their parishes in the shanty towns or slums. We discussed ways of helping the poor. We related the Bible to the reality in which we were living. We helped the poor with their sick children, and even with their household chores." In 1977, she and her husband left this work and joined the People's Army. The move was a personal reaction to the assassination of Father Rutilio Grande on 12 March 1977. He was gunned down as he drove to Sunday mass.

That year the military government of Colonel Arturo Molina had launched an all-out campaign to terrorize and kill parish priests and nuns. Anonymous pamphlets dropped into the street blamed the war on "Marxist priests." The slogan ran, "Be a patriot! Kill a priest!" When Rutilio Grande was killed, the newly appointed Archbishop Romero, a man noted for his moderation, openly condemned the Molina government. Molina had come to power in 1972 in a particularly scandalous election. Though his opponent, Napolean Duarte, was ahead at the polls by two to one, Molina's well-placed supporters in the previous administration stopped all election broadcasts and finally pronounced the colonel president. The Christian Democrats were outraged by the flagrant fraud. Molina's answer was to have Duarte arrested, imprisoned and tortured, although he was released when international pressure was brought to bear.

A month before his death Grande had told a crowd, "Nowadays it is dangerous and practically illegal to be an authentic Christian in Latin America. I greatly fear that very soon the Bible and the Gospel will not be allowed within the confines of our country. Only the bindings will arrive, nothing else, because all the pages are subversive — they are against sin."[2] Despite a government-declared state of siege, over a hundred thousand people risked their lives to attend Grande's funeral at a San Salvador cathedral. Eight bishops, Archibishop Romero and four hundred priests held mass for the slain father.

For Maria, who had known and loved the priest, it was a deep personal loss. It seemed that the last vestiges of humanity had been swept away, and in their place the war became uglier and uglier. That year when President Molina retired, his minister of defence, General Carlos Humberto Romero, was "elected" president in his place. The priest-killing campaign continued. Priests were found decapitated, disfigured by battery acid and otherwise mutilated. In Aguilares the army launched an attack in which several more Jesuits were murdered, and code-named it Operation Rutilio.

"Things really heated up," recalls Ileana. "Women in large numbers began to join the revolutionary movement. I think they found this final obscenity impossible to condone or ignore. Up until this time there had been a lot of groups in which the majority of members were women, but no women's organization per se. In 1977 the mothers of political prisoners and the disappeared formed an organization. Women from the markets also began to form groups. In 1978 the Association of Women in El Salvador [AMES] was formed in order to gather together all diverse groups so women could make more effective contributions and gain more power. It was set up to represent the needs and concerns of women."

Male chauvinism in Latin America has always burdened women there, especially in the lower classes, where it remained the only power left to unemployed, illiterate men, who dominated and made life even more miserable for their wives. In El Salvador the women have consistently been the labourers, working seasonally at harvest time. Often they have been abandoned with many children, their husbands leaving to spend time with other "wives" — sometimes for good, sometimes not. Without proper medical attention or adequate nutrition, one-tenth of all infants in El Salvador die either at birth or within the first year of life. It should come as no surprise that women, affected by social injustice in which their lot is inevitably worse than men's, have become a significant presence in the revolutionary forces.

Ileana: "We had some problems with chauvinism in the beginning. The men didn't want us to join, or they wanted us to stay in subservient roles. But soon they realized the importance of having as many people fighting as possible, and they changed a bit. I think it actually helped make male-female relationships more equal."

Maria adds, "I remember we talked about sexual discrimination in our meetings. It was decided that women should be allowed and encouraged to participate in all the activities — literature distribution, underground work, even fighting. Men were told they were equally capable of taking care of the children when their wives weren't there, and cooking and cleaning and washing. My husband did these things when we led a relatively normal existence. But usually we were far from being a regular family."

Ileana remained active in the "political" side of the front, but says there was not always a clearcut distinction between political and military. "We organized training practices in the countryside. I was taught self-defence, small arms, and how to make home-made explosives. I always carried a gun in the countryside because if the Security Forces came and searched the house I lived in and found my subversive literature, I would be killed. I had to be ready to defend myself."

Maria and her husband were asked to open a supermarket in San Salvador that would serve as a cover for shipping supplies to the guerrillas. "We lived over the store — my husband, our two daughters and two of my husband's sisters. Our house became a meeting place and the kids used to help camouflage it. When we were having a meeting one of us would go out and play with the kids on the street in front of the houses, making sure the coast was clear. We all carried guns for personal protection. We knew if we were searched by the authorities we would be killed. I used to carry a huge bag with all this kids' stuff in it — talcum powder, diapers, baby bottles. I hid my gun and pamphlets underneath. Sometimes I would even put my gun in my baby's diapers. The soldiers never thought of checking in the baby's diapers.

"The supermarket was a good front. We had a delivery van and I used to deliver goods to the groups in the countryside. I took them everything — shoes, beans and Kotex. Kotex was used for dressing wounds because if we took real bandages it would be too obvious.

"On one occasion when I drove to the place where I was supposed to meet our contacts, I was met by two members of the Civil Defence. I pushed the kids down on the floor of the van, jumped out and began running into the bushes. They opened fire and I fired back at them. I don't know if I killed them or wounded them, but the firing stopped. After a little while I returned to the van and there was no sight of them. My kids were still huddled on the floor where I'd left them."

Maria and her husband participated in several bank robberies in the capital. She described one that ended in tragedy. "We spent two weeks going over the plans down to the last detail — who would do this, who would do that; who would go in and give the order, who would stake out outside. It was perfectly worked out, but as the date got closer we got really nervous. The night before we did not even leave our home. We stayed in and drank lots of herbal tea to try to calm down. I took a couple of tranquillizers for my nerves.

"The next morning as we set out I was still really scared, but now it was a matter of self-discipline. We had to time everything, but when we drove up outside the bank things went wrong. Unexpectedly, the three policemen inside the bank were just being relieved and there was a patrol car waiting for them. One of our *campañeros* went in and demanded the money. He didn't realize there were now six cops in the vicinity of the bank, because he had come in from a different door. We couldn't get to him in time to warn him. I was staked outside, ready to cover the retreat. Suddenly I heard a lot of shooting and screaming. I rushed in and saw that three of my friends had been shot. Two were killed. The other, a woman, had been shot in the legs. I ran to her to help her get out. She said, 'Get out of here, Maria. It is better they kill three of us than all six.' She knew she would be killed anyhow, so she began screaming slogans. The police shot her to stop her from yelling out FMLN chants. I was running. I couldn't believe what was happening. It was so terrible to hear her screaming out and not to be able to help her. . . .

"My husband and I were severely depressed for quite a long time afterwards. We tried to bolster each other's spirits. We told each other that death was a part of the process of revolution."

In July 1979 the Sandinista revolution in neighbouring Nicaragua succeeded in overthrowing dictator Anastasio Somoza. "When we saw them succeed," says Ileana, "it gave us hope. I really didn't believe they could succeed so quickly. We thought if they can do it, we can too. But on the other hand, the United States learned a lot from the Nicaraguan revolution. Now they are applying that knowledge to El Salvador, giving enormous aid to the government for the military. They began sending in military advisers."

In Washington the Carter administration was sore on the point of human rights abuses, perpetrated by the Romero regime in San Salvador, which were making it increasingly difficult to get military aid bills through Congress. Washington needed a more moderate government in El Salvador. On 15 October 1979 a military-civilian junta overthrew the ruling regime in a brief *coup d'état.* The junta was comprised of younger, more moderate officers, and a number of representatives of opposition parties were appointed to the cabinet. The new government immediately began land reforms designed to restore land to the peasants. The reforms were ill-fated from the start. Army troops sent to redistribute parcels of land took over haciendas, helped themselves to the goods, then systematically murdered peasants who came to claim their new plots. Pay-offs and threats protected the estates of the richest families.

One by one the more moderate civilian politicians were forced out of office by the military leaders who controlled the army. One such politician, Hector Dada Hirezi, wrote in his letter of resignation: "The facts are indisputable proof of the conclusion. We have been unable to stop the repression, and those who commit acts of repression in defiance of the junta's authority remain unpunished; the promised dialogue with the popular organizations has not come about; the possibilities of generating reforms supported by the people have retreated beyond our grasp."[3]

Six months after the new junta took power, the archbishop of San Salvador, Archbishop Oscar Amulfo Romero y Galdames, was assassinated as he gave mass. *Time* and *Newsweek* magazines recounted the carnage that followed when thousands of people attended his funeral and army troops opened fire into the crowds. Three weeks after the archbishop's assassination the United States government committed another $5.7 million in military aid to the ruling junta. By this time it was estimated that in the country of five million people, two thousand people a week were dying in the war.

Ileana was caught in 1980. She says, "I can't tell you everything, but I was at the house of one of our *campañeros* for a meeting. The army found out about us and came and surrounded the house. We heard the trucks and jeeps pull up, and out stormed dozens of soldiers with machine-guns. We considered holding them off and trying to escape, but when we realized we were surrounded we surrendered. They arrested everyone in the house.

"I was taken to the National Guard's secret jail, where they interrogate political prisoners, and kept there for one week. I was raped repeatedly and tortured with electric shocks. I was three months pregnant, but thank God it didn't show. I knew if they found out it would be worse. They would have tried to hurt the baby, to abort it or something. They would have asked me who the father was. They continually threatened to kill me and my family. Sometimes I had to answer them, but I would just tell them things that they already knew, like where I had studied. Other times I would make up stories. I always thought about what I was saying and tried not to endanger the others.

"They didn't treat me better because I was a woman. To them there was no difference. I think for women it was worse. They thought we were worthless, so they wanted to defile us. I was constantly pawed, threatened with rape or raped. They were pigs." She pauses, then adds, "I know this sounds hard to believe. I was very lucky because many, many people never get out of those clandestine jails. I tried to keep myself together by telling myself how many others were in the same situation I was in. It made me stronger. When I was being tortured I kept thinking of my friends who had gone through this, as an example to keep me brave. I kept thinking that they had held out in even worse situations. It wasn't faith in God that kept me going. By this time I had lost my religious faith.

"I think in the beginning, despite enquiries from outside, the National Guard denied they were holding us. But the organization knew where we were. They notified our parents and began to lobby and talk to foreign diplomats and journalists and to people in the government. The archbishop even mentioned our names in his mass — the Church still had some influence. When the pressure from publicity grew, the National Guard handed us over to the legal institutions for a trial and prison sentence.

"We were very lucky," she admits. "Today we have so many people who have disappeared in the clandestine jails, and been killed. Despite pressure on the government, in the Church, no one reacts much to it now. Even international pressure and Amnesty International no longer have much power.

"We were taken to court. They said a lot of things. They said we were guerrillas, that they had found guns and subversive literature in the house, and that they had all the evidence. We were guerrillas, but the funny thing is that our house had been clean. They had no evidence at all.

"My husband didn't know I had been arrested until I was sentenced and sent to the penal institution. Of course he couldn't come to see me. But it was a regular prison run by the

Ministry of Justice, so I had visitors every Sunday. My family came and brought me some things — milk, because I was pregnant. I was kept there for four months."

On 5 March 1980 Napolean Duarte, the Christian Democrat who had run against Colonel Arturo Molina in 1972, been arrested and later exiled, agreed to head the eroding junta. Those who considered Duarte a moderate could not understand why he chose to become a part of the corrupt and brutal government, but those who knew him well have said that his tremendous ego eventually dictated his quest for leadership, even at the helm of a mutinous group of military leaders.

One of Duarte's first public statements was that "the Security Forces had been trained for fifty years to do things 'the other way.'" He said it would take "time to change things." He refused to negotiate with the FMLN despite the fact that Mexico, several European governments and many Salvadoran clergy recommended that he do so. Instead, he declared the country in a state of siege.

By 1980 the revolutionary forces had greatly expanded and were said to have gained widespread public support. They claimed in that year that forty percent of their leadership were women, and women were increasingly adopting military roles in the war. A women's military school was opened, offering a twenty-day training course to all women between sixteen and twenty-two years of age. At least two all-women battalions were formed, and Maria fought with one of them for a few months in the guerrilla-controlled zones of the countryside.

"In October 1980 I was sent to the liberated zone to prepare for the general insurrection we had planned. I needed more military training." The guerrilla camps were located as close as twenty miles from San Salvador. Most of them were hidden in the hills and jungles but could be reached on foot from the government offices in the city. When Maria arrived she noted a community atmosphere. Peasant women cooked and tended livestock that wandered around outside the assortment of mud and thatched-roof huts. Outside in the sun teenagers as young as thirteen and fourteen were taking apart their guns to clean them while they listened to rock music. At the camp Maria was taught to make explosives and to operate a G-3, German-made automatic rifle. Her husband had by now been in charge of several armed actions.

Against the army's superior fire-power, deployed in daytime search-and-destroy missions supported by A-37 Dragonfly jets and Huey helicopters, the FMLN pitted classic guerrilla tactics. The guerrillas continually launched lightning attacks and ambushes on the army patrols that passed through the region. At night they attacked army bases, throwing explosives into the barracks and shooting the soldiers as they tried to escape. They moved into small towns, only to vacate them again as soon as army forces appeared, operating on classic "everywhere and nowhere" guerrilla war tactics. Local peasants who support them, called *las masas*, passed on food and intelligence.

Maria participated in two main military actions during her stay with the guerrillas. "We planned to ambush an army patrol," she remembers. "We had to hide alongside the road and wait for the patrol to come by. I was terrified. I was clutching my G-3 and my hands were sweating, making it slippery. I felt like vomiting. As soon as I heard the first sounds of their armoured cars I sprang up and started to run away. I was thinking, 'I have two children, I'm going to get killed. . . .' One of the others grabbed my arm and said, 'You have to stay here. If you start running they will see you and start shooting. If you run, you are dead.'" Maria stayed and opened fire. "A bullet grazed my hand. Maybe that brought me back to reality. I wasn't frightened anymore."

Maria and her husband had left their two small daughters with friends in the city during this period. The couple spent time at several guerrilla camps, and finally Maria was sent to train for one month at an all-women camp of the women's Anti-Yankee Battalion. The battalion, consisting of about 250 women, operated in the San Cuentes area. Most of the women combatants were extremely young. "To be thirty in any of the guerrilla units was considered old," she smiles.

The women taught Maria many new manœuvres. "For example, how to cross a river using ropes. We would throw the ropes like lassos, and swing ourselves across the river holding the rope with one arm and our grenades and weapons with the other. We covered each other as we made our way. We left these ropes on the trees and periodically went to check to make sure they were not rotting and the knots were still secure. Work was the same as in the mixed guerrilla groups,'' she says. ''But when the enemy was killed or ambushed by the women's battalions they found it more demoralizing. They consider women worthless.

''We used this to our advantage. Whenever we successfully killed a number of army troops we always put out communiqués saying that we were responsible. We wanted to rub it in.

''I found that the all-women battalion was even more disciplined than the mixed units. For example, if we were given an order not to smoke all day because we were staked out somewhere, we wouldn't. If it had been men, someone would have found a way to sneak a cigarette. Women were more punctual about meeting times and places, too.''

Maria says that despite their youth most of the women combatants had children, who were left in care of their families or friends. ''Women still did most things during their pregnancies,'' she says. ''It was just for a short time after the baby was born that they couldn't do all the things they normally did.'' She shows an easy acceptance of motherhood, at any age or in any circumstances, that ''is prevalent in El Salvador. It just doesn't seem onerous to us to have babies. We don't wait for the right time and place.'' Despite the availability of birth control, both Maria and Ileana said that most women wanted to become pregnant because it was psychologically uplifting to give birth when so much death was going on around them.

Ileana had been released from jail and had given birth to a daughter. Six months later her husband was killed. ''They came at five in the morning to murder him,'' she says. ''That is when the death squads operate. Dawn is the most terrifying time for all of us. We would lie awake and half expect to hear the loud knocks and yells at the door. I hid myself and the baby while he went to hold them off. They took him away and shot him. Members of the organization came for me and took me to a safe place. I stayed in El Salvador for seven more months before they got me out to Canada.''

The FMLN planned to instigate a general offensive in January 1981 for which Maria and her husband had been training in the countryside's "liberated zones." They were told to return to the city in December 1980 to help co-ordinate the uprising there. On the first of December four American Catholic nuns were assassinated by members of the government forces. American President Jimmy Carter condemned the murders and ordered all military and economic aid to the junta suspended. Five weeks later, on 3 January 1981, two American economic aid consultants with the American Institute for Free Labor Development (AIRFLD) were shot to death in a hotel coffee shop, presumably by government enforcers of the agrarian "reform" program: the Americans had been privy to information that exposed corruption within the program. Washington leaders were outraged at the excesses, which were causing an uproar at home and making it increasingly difficult to get public support for American aid to the regime.

On 10 January Maria and her husband waited at home for their final orders. The next day guerrillas overran a classical music radio station and broadcast an appeal to the people of the country to rise up in a general insurrection.

A number of guerrilla units operating in the capital launched hit-and-run attacks against police and military targets. Maria was assigned to a unit attacking the air force base. "The base is located a little way from the centre of the city. We were to go in and plant the explosives, liquidate the guards and get as many of their munitions as we could. I was in charge of distributing arms to different groups after we got hold of them.

"The offensive began at seven o'clock in the evening, when there would be a minimum of patrol cars, and the majority of people would be at home. That way there would be fewer civilian casualties. We managed to attack the air base and get the weapons. But we were identified. They saw our car and the licence plate number. We returned home and hid the armaments in the back room."

The general insurrection had failed. The army had mobilized within the capital and imposed martial law and a dusk-to-dawn curfew. The FMLN had poorly co-ordinated their own forces and had counted, unrealistically it seemed, on the majority of the people in the city to rise up. In the countryside guerrilla gains were substantial, but in the cities the uprising was a failure.

The following morning the army surrounded Maria's home. "We heard their tanks and trucks and patrol cars surrounding the house. The trucks were full of soldiers. There were six adults and six children there, and we were all captured.

"I was taken to a cell where I was raped and beaten. For a few days they kept me inside a gas drum, and when they took me out for interrogation I would be tied hands and feet on metal bars, suspended horizontally with bags of sand on my stomach and then beaten. They used psychological torture as well. They would bring in my children, point pistols at their heads and ask me to talk. . . . I would make up things to trick them for a while.

"I was being kept at the National Police headquarters. In the beginning they said that they had documents to back up all their accusations. They told me they had spoken to my parents and that my parents had told them to kill me, had said bitter things about me, that I was a terrorist and should be killed.

"I was given electrical shocks, attached to all parts of my body. I was raped many times. I had tried to prepare psychologically for this. I had answers ready for them, a lot of garbage that they already knew or that was fabricated."

After two and a half months at the National Guard headquarters she was removed to Ilopango Prison for women political prisoners. Her children were allowed into the custody of relatives, her husband imprisoned elsewhere. The cell at Ilopango was her world for the next two years, during which time she participated in three hunger strikes that badly affected her health. The electric shocks had caused partial paralysis in her legs, which she is only now recovering from.

"Our capture was broadcast on the Liberation Radio, a station run by the FMLN. The archbishop, the Red Cross and Amnesty International began to lobby for our release. Because of the six children who had been taken at the same time there was a lot of across-the-board pressure on the government not to let us "disappear" — not to kill us. That is the only reason I am still alive. In April 1983 I was released from prison. My husband was already dead — he was killed at some point during those two years. I left my children with my relatives and came to Canada. I was sent here by the FMLN to recuperate, basically. This is a recovery period."

Since Maria has been in Canada, the civil war has escalated in El Salvador. The Reagan administration continues to pour large amounts of military aid into the country despite continued pressure and documentation, by human rights organizations, of the junta's atrocities. The killings continue, indiscriminately; labourers, priests and nuns, students, suspected leftists and even foreign journalists are targets. The majority of the killing is attributed to the right-wing death squads operating for Napolean Duarte's government. He has ruled continuously since 1980 and was "elected" to office in June 1984 when he ran against an ultra-rightist candidate without any challenge from FMLN candidates. As one foreign journalist wrote: "The government's stand — and Duarte's — is that the guerrillas must lay down their guns and join an electoral process set up by the government. For the guerrilla movement, the problem with this stance is that it may mean both literal and figurative suicide. Literal because leftists could not campaign in El Salvador without getting murdered. . . . The left could never afford to lose an election, and the U.S. government would never allow it to win."[4]

The FMLN has moved noticeably further to the left and undoubtedly receives financial support from the Socialist bloc. One American journalist who travelled deep into guerrilla zones recently was asked by local villagers if he was the guerrillas' Russian adviser. It is apparent that despite the many political beliefs that united to form the FMLN, it has now turned for help to those who are sympathetic — typically the Socialist countries. Once again a civil war fought only because of brutal internal oppression has become a battleground for the superpowers. Still, women like Ileana and Maria remain convinced that the only way to effect change in their country is to fight.

"It is such a strange state of consciousness to leave that reality and come to this one," says Maria. "I don't think I have really adapted at all. There are always reminders. I am never without the presence of El Salvador. If I am eating a good meal,

I think about those who don't have enough to eat there, people so poor that they eat roots and weeds. I think of my children every minute. It has not been a tranquil time here, either.

"So little is brought out in the press here about that reality. The reasons for the war, the things that moved people like myself to fight. There are a lot of places where the children only eat once a day, or don't eat at all some days. These are terrible facts that aren't being discussed. If people think that the war in El Salvador is a Moscow- or Cuba-directed Marxist revolution, they should come down and spend some time in our shanty towns, with our peasants or in our prisons. Maybe then they would understand why we fight. We take help where we find it. If the Americans had decided to help the people overthrow fifty years of military rule, no one would call this a Marxist revolution. But instead, they support a government responsible for the disappearance of ten thousand people each year. They forced us to turn elsewhere for help. . . .

"I'm ready to go back. That is not a difficult decision. My first concern is to get my children out of there and bring them safely to Canada. Then I will find a sympathetic family to take care of them, and I will return. I know it will be difficult to prepare to go back, but I will go. I couldn't remain living here, knowing what is going on there. I would hate myself for not continuing to fight for the things I believe are necessary."

"It is an area of strategic importance to the United States," Ileana told me before she returned, "so they continue to support the government and ignore the human rights abuses. I think the war will continue for a long time. Mothers are fighting, kids are fighting, priests and nuns are fighting or helping us to fight. We won't give up. We have tried to negotiate, but it never works. It is not to their advantage to come to an agreement with us. They have all the power and all the wealth, so why should they? They just go on suppressing the movement and increasing their military strength."

In late 1984 Napolean Duarte arranged the first meeting to discuss future negotiations between his government and leaders of the FMLN. The meeting was positive, if only in bringing the two sides together, and all agreed to meet again.

Meanwhile war and rumours of war escalate throughout

the region. The Sandinista government of Nicaragua fears an attack from the United States, whose government has accused them of disrupting the balance of power in Central America and arming the revolutionaries in El Salvador.

Maria: "We are willing to die to change the basis of our society, to feed and educate our children, to end the murders, the terror and the oppression. You might not understand that, here. But there, it is another reality."

EPILOGUE:
War and Peace

War changed me. You cannot go through that and come back the same. I've been to a place where nothing is sacred except what you have inside. What keeps you going is that little inner part that you own and that is really all you do own. . . .

Lynn Bower

Two of the women in this book are still at war, but for the others there have been years of peace since the events they describe. The intervening years have meant readjustment, for some a whole new life and country. They have brought many changes, social and personal. For most, they have been years of healing. My final question to each woman was twofold: How did war change you, and what are your feelings about war and the preservation of peace?

More than anything else, there was a sense of lost youth in their final reflections. We forget that war is fought by the very young. These women, like any man who goes to war, all lost something of their innocence in war; they aged beyond their years. As members of generations whose lives were, for a time, given to war, they will never be quite the same as the rest of us.

Diana Barnato Walker never remarried. She directed her energies to the care of her beautiful country home, her sheep and horses. Her house and life seem to defy change. "The war made me sad," she reflects. "Now I try never to think about it. I think my philosophy and outlook were altered by the people I met whom I never would have met otherwise. The privations we went through physically and mentally certainly altered us. Priorities have changed — what is important. Losing loved ones and hearing about death affected me greatly. It has left a mark."

Joan Cowey came to a new life and a new country after the war. She has remained married and has eight children and several grandchildren. "The war prepared me for life. My life hasn't been that easy. I've had nine children and lost one. I've survived quite a bit, and yet I don't panic easily. The war prepared me for sorrows and hardships. . . .

"I think about all my army friends and really miss them. It was one of the best times of my life. Everybody helped everybody else. It was exciting. I would have had a boring life if I hadn't joined the army. That is the way I feel, even though I saw the awful side of war. I realized what war was really like, but it was exciting for a young person who didn't give a hang, really.

"Today I get very rattled when I watch the news and they say things like, 'For the first time a woman had done such-and-such.' Holy smokes, we did all those things forty years ago. They just ignored us, that is what it really was. In regards to women . . . I think there will always be ordinary wars, and women will always be needed. I'm sure women will be allowed in combat."

Jeanne Bohec kept busy as a single parent raising her son and worked as a teacher in Paris. Today she is a grandmother. "I can't say how the war changed me, exactly. Before it I was a young girl, and afterwards a woman. But certainly the war has affected the rest of my life, because I had this image: I was a woman who had done 'that sort of thing' and now had a reputation. People don't treat me like a feeble woman. Men treat me as a comrade. One friend of mine, a general, recently called me a 'sacre bon homme' — a very good fellow. I said, 'No. I am a very good woman!'

"They were exciting days, but when it was over it was over. The men miss it, I think. They always talk about their experiences and we call them 'anciens combattants' [old soldiers]. But women talk about it rarely. Today when I have to speak about it I do it simply. Most women talk about it even less than I do. Many times I have been bored by men who talk endlessly about the war. I never hear women talking about it.

"I believe war is an awful thing. I believe we should not have it but if we must defend our country, I think that women, like men, should fight — except those who have young babies. Women can learn everything they need to know about fighting. We should do it together."

Marisa Musu's life has remained full and intense, with marriage, children and a career in journalism. "How did war change me? In a sense it gave me a great amount of self-confidence — not because of my experience as a partisan, but when I was arrested and knew I was going to be shot the next morning. In those few hours I decided whether or not I would betray others and thereby save my life. Once you have that confidence you behave with conviction. There are some women who miss the adventure, but not me. For many there is a sort of nostalgia and regret for not having found anything else as interesting or challenging or exciting to do. I know many people for whom those ten months of resistance in youth spoiled their entire lives because they have continued to live in their memories of those experiences. The important thing is to consider these activities as something just and well done and leave it at that. As a matter of fact, my children have not learned about these things from me. As for how women look at war . . . perhaps women have more feelings. I don't know, I'm not sure."

Carla Capponi has remained active throughout her life, fighting for causes to which she is deeply committed. She raised her daughter, Elena, as a single parent and endured the debilitating surgery and loss of one lung, seven years in a sanatorium and a lot of hostility directed towards her for her role in Via Rasella. "The war made me more responsible. I became involved in politics right afterwards. I have been greatly involved in political activities especially as related to helping women. I have been city councillor in Rome and have organized many protest demonstrations to obtain some changes in the city such as pulling down unsanitary housing, and other projects. The war did not diminish my moral strength. It's not as though I went back to domestic life. Yes, many women did become housewives, but many also entered politics or pursued professional careers. Each of us has had civic and political responsibilities. So you can see that we have continued to fight. We found that there was a connection between what we were doing during the war and what we fought for after: wanting to change our country into a more civilized, freer, more democratic nation where there would be great social justice."

Ida Kasprzak today lives with her husband in Canada. Her war changed not only her country of habitat but also "the way I look at things. Now I believe in reincarnation. Why else did so many people accurately predict their deaths? Why else did my instincts tell me when it was 'a bad day'? These things imply more than just animal instincts — a strong psychic connection that does not fail to give warning and protection.

"I think what women did in all countries during the Second World War was remarkable. War is not a man's domain. It can't be anymore.

"I hate war. I only want peace. But I would fight again to defend Canada and Poland, my two countries. I would fight to defend liberty. I don't believe in peace at any price."

Irena (Black Barbara) Komorowski married a Polish soldier and raised their son in their new home in Canada while working in a trauma ward at Toronto General Hospital: "I always had nightmares . . . lots of nightmares, but I kept my good nature. I laugh a lot and I sing. But I no longer believe as much in the goodness of mankind, because I have seen such terrible things . . . terrible things.

"I'm not an activist, but inside I am a pacifist. After the war I served two years in Germany. I saw the poverty, hunger, complete devastation. I had the feeling that we were finally even and we should forget about it. Gradually I became disinterested in warmongering. I remain an optimist about nuclear war, because I don't think either the United States or Russia wants a war like that. It was a bloody and terrible war in most European countries and we all suffered. Only somebody absolutely sick or crazy would want a nuclear war.

"My memories are still strong. Even a smell or a certain kind of lighting will immediately bring back images, memories. But gradually they do fade a little. I tried to raise my son to know about these things, but I think there is a natural rejection. The young don't want to hear. It is our story, not theirs."

Nadya Popova today remains married to the pilot she met

during the war. She is a member of the Moscow Veterans' Association, busy with its administration, and herself a grandmother. "Of course those years left a definite imprint on me. I can speak not only for myself but for my generation. I can speak on behalf of my friends, both men and women. Those years left an imprint on us all. They reaffirmed the awareness of the meaning of peace and friendship. They reaffirmed the awareness of the great joy of simply living — to value every day of one's life. If you look at my generation most of them are already elderly or crippled, but they still want to be useful. When they talk to young people they try to persuade them what a great happiness it is simply to live. And we must try to do everything we can to ensure that such a tragedy will never happen again.

"But despite everything I saw and experienced, I have remained an optimist. I was always cheerful, buoyant, as a child, and I remain so, notwithstanding the fact that I am sixty-four years old and have lived a long life. I have preserved my curiosity about life. I want to know, to see more, to accomplish more. Of course it left psychological scars: I can still see the images of friends crashing and dying in front of my eyes. . . .

"I have visited many countries and met people from all walks of life, from ordinary people to kings, and I believe that all people want to live in peace, no matter what country they are in or what their position in life is."

Sophia Kuntsevitch lives today in Minsk, Byelorussia, and still works in education. She too is a grandmother. "The war changed my life. I gave up the chance to be a doctor because of the graveness of my wounds. I did not have the strength left, so I went into the faculty of library sciences instead. How did I change? I think I became a little more introspective, a little more inside myself."

Katyusha Mikhaylova became a doctor, married and had children and today lives in Moscow. She is an active member of the Women's Association for Peace. "How did I change? Later my patients told me that, as a doctor, I was very humane. They all accepted my treatment readily, and they liked me a lot.

"We women must give birth to children, take care of them, love and not fight. We don't need war. We hate war. War is unnatural. War kills people. Now we have to defend peace with all our strength at any cost. We need peace — particularly with the threat of nuclear war. I know that not only would all mankind die in such a war, but our planet would be destroyed. As a woman who fought in war and who knows its price, I ask all young people to fight for peace. It is your future. . . .

"The Americans have called our war the Unknown War. It was not unknown to me. I was wounded in my arm, my leg, my stomach. I still have twenty-two bullet fragments in my body. I know this war. I tell my children about it often. I want them to know about it and not forget it, ever. I believe that if we tell the young about war, they will defend peace. If we were ever invaded again, I would fight. But like all women I prefer to raise my grandchildren and live in peace."

Hanna Armoni today lives in Tel Aviv and works in the Historical Museum of the Freedom Fighters of Israel (Lehi). She has raised three daughters and is now a grandmother. "I accept reality. Reality is not just the best side of life, or the worst — I think that life is a package deal of sorrow and happiness. You have to take it together and know how to keep the balance. But I have never been free of the memories of those who were killed. I carry them all in my heart. I carry my sorrow. . . .

"I would do it again. I still believe it was the right way and the only way. It was the most important time of my life."

Shoula is also a grandmother today. Her children, still in Israel, have all served their time in the Israeli defences: "I hoped for my children that there would be no more war, but it didn't happen that way. Maybe my father hoped the same for me, and his father for him. War is terrible, and yet for Israel it has become a way of survival. Things happen so fast in wartime that you don't have time to think about it. Some wars take place in a matter of months. I think about that a lot. It's only later, or sometimes before, that you can really decide what you do think about it. At the time the momentum — everything — you just do what you have to do.

"But I'll tell you something, now it's another kind of war. There must not be battles anymore. We can push a button and everything blows up. I hate war. Maybe if we all said how much we hated war it would lead to peace for all of us. I wish it were true, but that is utopian. But now, with modern technology, we must stop wars."

Yaffa married an American and moved to the United States, where they have raised their family. She remains, however, an Israeli citizen, and a very patriotic one. "The war made me stronger. It was the most important experience of my life. Today I am against anything nuclear, even nuclear energy. It is too dangerous. . . .

"On the other hand, I am not a pacifist. I don't think you can be a pacifist in Israel. We have to have a strong defence — we have to defend ourselves there. As for my other country, the United States — I believe in democracy, and fighting against terrorism. I don't believe in interfering in other countries, in supporting dictatorships that push the people into the Communist camp. I believe we must help the people, not support brutal government just because they are pro-American.

"Finally, I believe strongly in women taking part in their country's defence, not because I am a women's libber but because I feel we have to share the responsibility. How can women send their husbands, their brothers and their sons off to fight, and just sit at home? War is terrible. I hate it. But if we are going to have it, I think it is the responsibility of all of us."

Shifra lives today in Toronto with her husband. She has recently become a grandmother. "My outlook on life is so hard to define at this stage. I can only tell you that I am what I am today because of my experiences thirty-five years ago. I'm obviously stronger. I cope better. With me it became positive. Someone else may have cracked under those circumstances, but I didn't. Still, I don't like to think about it. I did what I had to do. If it wasn't me it would have been someone else.

"No sane person likes war or wants to be embroiled in it. But as horrible as war and killing is, I still believe that our people had no alternative but to defend their very existence."

Geneviève de Galard still lives in her family apartment in Paris with her husband of almost thirty years, who has recently retired from the French army. "It has always been difficult for me to talk about it with people who were not there, especially those who are curious rather than sympathizing. I never regret being there, though. For me it was an honour.

"People have always been very generous towards me. Nurses are very respected and honoured in France, especially army nurses. I have had a very busy life, and for a long time I didn't want to speak about the war, except with those who had been there. I now speak to the young because I think it is important for them to know what happened. I am a living record of the courage I saw, and I feel I must tell people about it. I think nowadays the young need to have examples of courage and a sense of honour.

"War is something horrible. But I think that, faced with the deprivation of freedom, I would be ready to fight again. I wish for peace, but not at any price. Freedom is worth a fight."

Brigitte Friang is now working on her next book. She is often ill, as a result of her wound and the abuses of twenty-five years in combat zones, but she remains intensely active. Today she says: "In war I have seen human nature reveal itself. I've seen the best and worst in people. During war some people are able to kill and cause pain, although in civilian life they might be good people. On the other hand, some people do magnificent things during war to help a comrade or save someone's life. They become far better than they would ever have been in civilian life. It is a fascinating study. I hate war, but the experience has given my life remarkable friendships, and intensity.

"Ironically, war taught me absolute respect for life. I think of death constantly, at least ten or twenty times a day. I don't know what will happen when I die, or what I shall do . . . but I have faced it so many times, and I am not frightened. I think of each day as a gift, even now. The haunting of death has made me live life more fully."

The second generation of women who went to Vietnam are only just beginning to talk about it. Lynn Bower has continued nursing and is raising her son on her own. "Would I do it again? I would go just to keep one younger person away; to spare one younger nurse. But I believe that war doesn't solve anything. I am happy to see that the United States is finally setting up a Peace Institute. We should have had one since the day this country was founded. I am concerned, because my son is twelve years old, and I've had recurring dreams of him coming back to me in a body bag. I don't know how mothers can raise their children knowing that their countries might take them, for God's sake, at eighteen. At eighteen you haven't done anything yet. . . .

"I'd like to see women soldiers go into combat. I think that more women with children should be in, because I think if we sent more women into wars, wars would not last very long, because you cannot carry a child for nine months, nurture him for eighteen years and then see him slaughtered, without saying, 'Stop, it's enough, there is another way. . . .'

"War changed me. You cannot go through that and come back the same. I've been to a place where nothing is sacred except what you have inside. What keeps you going is that little inner part that you own and that is really all you do own. You have to take care of it — dignity."

Betty Stahl Doebbeling today continues work in a nursing home and is busy with family life. She married a veteran of the Korean War and raised a family. "It change me, I suppose. It was the most significant time of my life. I miss that time as many people do. I've tried to identify what I miss, but I don't know. Maybe the challenge, the importance of it. You saw all sides of life, people handling pain, stress, loneliness, all those things. And you tried to see if any of that was going on inside yourself.

"As for my thoughts about war, I have to go back to my spiritual beliefs. The Bible says there will be wars and rumours of wars for all time. I think we can speak until we are hoarse, but there will always be factions of people who will do anything to get what they want, and they are going to take what they want and will probably call it war. I certainly would be willing to fight against someone trying to take our liberties away."

Donna-Marie Boulay married a Vietnam veteran, gave up nursing and became a lawyer specializing in law concerning nursing. She is a member of the Lawyers' Alliance for Nuclear Arms Control. "You can't come back the same. I learned a tremendous amount, and that changed me. I felt a lot of anger, and still feel it, when it comes to old men waging war and young men paying the price. It is the young men who have to live the nightmare. I guess woman have the history and cultural expectations to be the preservers of life, to nurture it. We nurture our children, our siblings, our parents when they get old and our spouses. As a profession, nurses are expected to nurture. I think the more men also learn to do that, the better. It is one of our hopes for preventing the cruelties that happen, including war."

Mary has continued nursing, and after Vietnam spent two years in Bangladesh. "Now I am very anti-nuclear anything. I've seen what men can do in war — boys who were decent being thrown into a situation that was so mind-boggling that it distorted everything, including them. I would probably be in prison today for demonstrating against the arms race if my job hadn't put restrictions on me. But if we go to war again I will definitely speak out, even if I lose my job because of it. The experience in war prepared me for a lot of things in my life. I know it prepared me for Bangladesh."

Mary Beth Crowley still nurses at a veterans' hospital. "Today it seems that it is finally okay to be a woman veteran. It is finally all right to acknowledge you were there. I'm very happy about that.

"I believe it is getting to be my turn to protest nuclear arms. People who haven't been to war think more ideologically. When you get there you quickly lose those ideals and start dealing with the real thing — what actually goes on — and then it is very hard to go back to thinking the same way. And I do believe that women see and feel things differently. Women are touched differently. I think women have less of a need to prove themselves. Deep down, most of us know who we really are, and we don't have to be part of the group. So I can say I don't believe in fighting and guns and war."

Maggi Arriola has also continued nursing and is a single parent. "Vietnam is the most significant thing that has ever happened to me. It will always be there and will always affect how I look at things. Because of Vietnam, I feel it is very important to tell and show the people I am close to that I love them. I don't just save it up for a specific time or when we are together. Sometimes I feel like just calling up my brother and saying, 'Hi, I was thinking about you.' Before I would never have done that — I would have worried about the cost of a long distance call, or something like that. I appreciate people as individuals more. I question my religion more. I haven't lost my faith, but I have felt the need to examine it. . . .

"I would feel a lot more comfortable if we didn't have nuclear weapons, but I feel that they are here and we have to deal with them in a responsible manner. I've seen men die, and I sure don't want to see that happen to my son. If I were asked to go to another foreign country and I didn't understand why, I would definitely be more questioning. I would never go blindly just because I was asked to go. . . .

"I remember their faces all the time. I remember them in their fatigues. I've seen people exhausted, but I've never seen anyone since with that kind of look. I wonder if people can see that, when they watch the documentaries or look at the photographs?"

Linda Kitson's three months in the Falklands War became another year of work on her exhibit, followed by work on a film about the war in Cambodia. "In perspective, I think it changed me. I think now more about Vietnam and the scars that all those years left on millions of people. War is now constantly in my work, my thoughts, my sleep.

"I watch the news, see the controversy over cruise missiles and all that, and feel that never the twain shall meet. I've become cynical and convinced that although there should be nuclear disarmament, it will never get through in our lifetime. I was very confused about war before. Now, I'm afraid that I believe it is with us to stay, it is a part of life. You just can't wipe out human instincts that tell us to defend whatever we think is ours without a very slow, evolutionary genetic change in our make up. It only takes one man to stir up greed and tell us we want land, gain and, of course, political control. The whole thing has made me sadder and sadder."

War and rumours of war continue to rage throughout the world. There are children who grow up never knowing peace. Ileana and Maria are still living in what Maria called "another reality." For them there are no final thoughts about war because it is still *the* war, their war. They think only about survival and hope for the future, hopes for their children, that they may live their lives in peace.

NOTES

HOLDING THE LINE

1. Anne Valery, *The Passing-Out Parade: A Play* (London: Samuel French, 1980), author's notes.

2. Lettice Curtis, *The Forgotten Pilots: A Story of the Air Transport Auxiliary, 1939-45,* 2nd ed. (Twyford: E.L. Curtis, 1982), 11. The author quotes C.G. Grey, editor of a journal called *The Aeroplane.*

3. Ibid., 21.

4. Former ATA pilot Marion Orr, interview with author, Toronto, 1981. Five Canadian women flew for the ATA during the war. In addition, there were twenty-five American women pilots, three Poles, one Chilean and women from Australia, New Zealand, India, South Africa and Ceylon (Sri Lanka).

5. *The Times,* 21 Nov. 1941.

6. "Parliament Debates Conscription of Women," *The Times,* 11 Dec. 1941.

7. Ibid.

8. Ibid.

9. Author and former ATS member Anne Valery, interview with author, London, 1983.

10. Dame Helen Gwynne-Vaughan, *Service with the Army* (London: Hutchinson & Co., 1943), 101-2. General Gwynne-Vaughan helped create many of the ATS's early programs.

11. Valery, 1983.

12. Ibid.

13. Diary of Mari Hopkins, ATS woman. Imperial War Museum, London.

14. "Mr Churchill's Visit to AA Women," *The Times,* 18 Oct. 1941.

15. "ATS Girls Work on Gun Site," *The Times,* 11 Oct. 1941.

16. Orr, 1981.

17. Ibid.

18. Anonymous ATA woman, interview with author, 1981.
19. "They Shot 'Em Down," *Stars and Stripes*, 1942.
20. There is an excellent book about the American women pilots, the WASPs, who flew during the war: Sally Keil, *Those Wonderful Women with Their Flying Machines* (New York: Rawson Wade, 1979).
21. Nancy Loring Goldman and Karl Wiegand, *The Utilization of Women in Combat*, U.S. Army Technical Report 563 (unpublished), 1982, 99.
22. Diary of Anne Laws, ATS woman stationed in Belgium, 1944. Imperial War Museum, London.

LE RESEAU

1. M.R.D. Foot, *Resistance* (London: Methuen, 1976), 13.
2. Quoted in David Schoenbrun, *Soldiers of the Night: The Story of the French Resistance* (New York: Dutton, 1980), 40.
3. Jeanne Bohec, *La plastiqueuse à bicyclette* (Paris: Mercure de France, 1975), 81-82. Passages translated for this work by Debi Goodwin, Toronto.
4. Ibid., 81.
5. The first female SOE agent was sent into France by late May 1941.
6. Marie-Madeleine Fourcade, *Noah's Ark: The Secret Underground* (New York: Kensington, 1974).
7. Brigitte Friang, *Regarde-toi qui meurs*, 1: *L'ordre de la nuit* (Paris: Editions Robert Laffont, 1970), 35-38. Passages translated for this work by Debi Goodwin, Toronto.
8. Ibid., 39-40.
9. Bohec, *La plastiqueuse à bicyclette*, 96.
10. Friang, *Regarde-toi qui meurs*, 1:48.
11. Bohec, *La plastiqueuse à bicyclette*, 115.
12. Friang, *Regarde-toi qui meurs*, 1:54.
13. Bruce Marshall, *The White Rabbit* (London: Evans Bros., 1952), 95.
14. Friang, *Regarde-toi qui meurs*, 1:110-15.
15. Marshall, *The White Rabbit*, 140-41.

16. Yeo-Thomas was deported to Buchenwald concentration camp, from which he escaped. He was recaptured and taken to a French Stalag. Again he escaped, and this time he found his way to the American lines. For his story, read Bruce Marshall, *The White Rabbit*.
17. Friang, *Regarde-toi qui meurs*, 1:188.
18. Bohec, *La plastiqueuse à bicyclette*, 145-47.
19. Friang, *Regarde-toi qui meurs*, 1:169-71.
20. Claude Chambard, *The Maquis: A History of the French Resistance*, trans. Elaine P. Halperin (New York: Bobbs-Merrill, 1976), 146-53.
21. Bohec, *La plastiqueuse à bicyclette*, 209.
22. Friang, *Regarde-toi qui meurs*, 1:241.
23. Ibid., 232.
24. Ibid.
25. Ibid., 369-71.
26. Bohec, *La plastiqueuse à bicyclette*, 37-38.
27. André Malraux included Brigitte's testimony in his chapter "La condition humaine" in André Malraux, *Anti-Memoirs* (New York: Holt, Rinehart and Winston, 1967).

VIA RASELLA

1. All interviews for this chapter were translated by Marianne Gilbert, York University, Toronto. She informs me that this epitaph sounds much more poetic in Italian.
2. Rosario Bentivegna, *Achtung Banditen!* (Milan: Ugo Mursia Editore, 1983), 135-36. Passages translated for this work by Marianne Gilbert.
3. Ibid., 139.
4. For a detailed account of the Via Rasella episode, read Robert Katz, *Death in Rome* (New York: Macmillan, 1967). Mr Katz interviewed a number of the participants and other principal characters for his work.
5. Pope Pius XII has been accused of failing in his Christian duties during the Second World War. He did not publicly protest the murders of millions of Jews, or even intervene against the deportation of a

thousand Roman Jews to their deaths in 1943. Further, he did not excommunicate the Nazi murderers, though he did later excommunicate Communist leaders. Some historians are kinder, claiming he was humane but cautious.

6. Peter Tompkins, *A Spy in Rome* (New York: Simon & Schuster, 1967).
7. Bentivegna, *Achtung Banditen!*
8. "Italy," in *Women in Combat and as Military Leaders,* U.S. Center of Military History Report (unpublished), 1978.
9. Katz, *Death in Rome,* 238.

UPRISING

1. See the next chapter, "The Eastern Front," for details.
2. Ida published a short account of her experiences in the *Polish Combatants Association Quarterly* (Toronto), July 1984.
3. J.K. Zawodny, *Nothing but Honour: The Story of the Warsaw Uprising* (Stanford, Calif.: Hoover Institute Press, 1978). The author makes several references to Black Barbara.
4. Ibid., 38.

THE EASTERN FRONT

1. K.J. Cottam, ed. and trans., *The Golden-tressed Soldier* (Manhattan, Kansas: MA/AH, 1983), xi.
2. For more details on women pilots training at Engels, see K.J. Cottam, *Soviet Airwomen in Combat in World War II* (Manhattan, Kansas: MA/AH, 1983), and Bruce Myles, *Night Witches: The Untold Story of Soviet Women in Combat* (Novato, Calif.: Presidio Press, 1981).
3. Tatiana Tsherbinovskaia, interview with author, Moscow, May 1984.
4. Clara Tickondvitch, interview with author, Minsk, May 1984.
5. Ibid.
6. Cottam, *The Golden-tressed Soldier,* "Introduction."
7. Ibid., 243-55.

ENDINGS AND BEGINNINGS

1. Peter Calvocoressi and Guy Wint, *Total War: Causes and Courses of the Second World War* (Harmondsworth: Penguin, 1972), 556.
2. Yigal Allon, *The Shield of David: The Story of Israel's Armed Forces* (London: Weidenfeld & Nicolson, 1970), 128-29.
3. Ibid.

AN EAGLE AND AN ANGEL

1. Bernard Fall, *Hell in a Very Small Place: The Siege of Dien Bien Phu* (Philadelphia: Lippincott, 1967), 17.
2. Ibid., 74.
3. Brigitte Friang, *Regarde-toi qui meurs, 2: La guerre n'a pas de fin* (Paris: Editions Robert Laffont, 1970), 264.
4. Ibid., 59.
5. Ibid., 65-66.
6. Ibid.
7. Ibid., 52-53.
8. Fall, *Hell,* 75.
9. Ibid.
10. Friang, *Regarde-toi qui meurs,* 2:85-86.
11. Ibid.
12. Paul Grauwin, *Doctor at Dien Bien Phu* (New York: John Day, 1955).
13. Ibid., 140.
14. Ibid.
15. Fall, *Hell,* 190.
16. Grauwin, *Doctor at Dien Bien Phu,* 150.
17. Ibid., 303.
18. Letter in private collection Geneviève de Galard de Heaulme de Boutsocq, Paris.
19. "Fortress Nurse Returns to Hanoi," *New York Times,* 24 May 1954.
20. "Madame Parachute," *Newsweek,* 13 May 1954.
21. Friang, *Regarde-toi qui meurs,* 2:263.

TWILIGHT ZONE

1. Lynda Van Devanter, *Home before Morning: The Story of an Army Nurse in Vietnam* (New York: Beaufort Books, 1983).

2. The story of Vietnamese women in thirty years of war, 1945-75, is truly remarkable. According to U.S. intelligence estimates in 1968, half the members of many guerrilla units were women fighters. That same year the North Vietnamese government reported that 250,000 women fighters had been killed, more than 40,000 disabled, and 36,000 were POWs. Despite efforts, I was unable to get to Vietnam to interview some of these fascinating women.

3. Donna-Marie Boulay, "No More War," *The Kabekona Journal* (Laporte, Minn.), 1982.

LASTING IMPRESSIONS

1. Dame Elizabeth Frink, "Foreword," in *The Falklands War: A Visual Diary,* by Linda Kitson (London: Mitchell Beazley and The Imperial War Museum, 1982).

2. Linda Kitson, *The Falklands War: A Visual Diary* (London: Mitchell Beazley and The Imperial War Museum, 1982), "Introduction."

3. Ibid.

4. Ibid., 37.

5. Ibid., 39.

6. Ibid., 85.

7. Ibid., 71.

8. Linda Kitson, interview in *The Times,* 4 Aug. 1982.

9. Kitson, *The Falklands War,* 102.

ANOTHER REALITY

1. Penny Lernoux, *Cry of the People* (Harmondsworth: Penguin, 1980), 70.

2. Ibid., 74.

3. Quoted in Liisa North, *Bitter Grounds: Roots of Revolt in El Salvador* (Toronto: Between the Lines, 1981), 82.

4. Guy Gugliotta, "Duarte's Second Chance," *The New Republic,* 13 Aug. 1984.

BIBLIOGRAPHY

Ales, Adamovich. *Out of the Fire.* Moscow: Progress Publishers, 1980.

Allon, Yigal. *The Shield of David: The Story of Israel's Armed Forces.* London: Weidenfeld & Nicolson, 1970.

Asprey, Robert B. *War in the Shadows: The Guerrilla in History.* Garden City, N.Y.: Doubleday, 1975.

Battaglia, Roberto. *Story of the Italian Resistance.* London: Odhams Press, 1957.

Bentivegna, Rosario. *Achtung Banditen!* Milan: Ugo Mursia Editore, 1983.

Binkin, M., and Bach, J.S. *Women and the Military.* Washington, D.C.: Brookings Institute, 1977.

Bohec, Jeanne. *La plastiqueuse à bicyclette.* Paris: Mercure de France, 1975.

Bruce, George. *The Warsaw Uprising.* London: Rupert Hart-Davis, 1972.

Calvocoressi, P., and Wint, G. *Total War: Causes and Courses of the Second World War.* Harmondsworth: Penguin, 1972.

Chambard, Claude. *The Maquis: A History of the French Resistance Movement.* Translated by Elaine P. Halperin. New York: Bobbs-Merrill, 1976.

Cottam, K.J., ed. and trans. *The Golden-tressed Soldier.* Manhattan, Kansas: MA/AH, 1983.

———. *In the Sky above the Front.* Manhattan, Kansas: MA/AH, 1983.

———. *Soviet Airwomen in Combat in World War II.* Manhattan, Kansas: MA/AH, 1983.

Curtis, Lettice. *The Forgotten Pilots: A Story of the Air Transport Auxiliary, 1939-45.* 2nd ed. Twyford: E.L. Curtis, 1982.

Fall, Bernard. *Hell in a Very Small Place: The Siege of Dien Bien Phu.* Philadelphia: Lippincott, 1967.

Foot, Michael Richard Daniel. *SOE in France.* London: H.M. Stationery Office, 1966.

———. *Resistance.* London: Methuen, 1976.

Fourcade, Marie-Madeleine. *Noah's Ark: The Secret Underground.* New York: Kensington, 1974.

Friang, Brigitte Elizabeth. *Regarde-toi qui meurs.* 1. *L'ordre de la nuit.* Paris: Editions Robert Laffont, 1970.

———. *Regarde-toi qui meurs.* 2. *La guerre n'a pas de fin.* Paris: Editions Robert Laffont, 1970.

Gettleman, M.; Lacefield, P.; Menashe, L.; Mermelstein, D.; and Radosh, R., eds. *El Salvador: Central America in the New Cold War.* New York: Grove Press, 1981.

Goldman, Nancy Loring, and Wiegand, Karl. *The Utilization of Women in Combat.* U.S. Army Technical Report 563 (unpublished). Research Institute for the Behavioral and Social Sciences, 1982.

Grauwin, Paul. *Doctor at Dien Bien Phu.* New York: John Day, 1955.

Gwynne-Vaughan, Dame Helen. *Service with the Army.* London: Hutchinson & Co., 1943.

Katz, Robert. *Death in Rome.* New York: Macmillan, 1967.

Kitson, Linda. *The Falklands War: A Visual Diary.* London: Mitchell Beazley and The Imperial War Museum, 1982.

Knightley, Philip. *The First Casualty.* London: André Deutsch, 1975.

Korbonski, Stephan. *The Polish Underground State 1939-45.* New York: Columbia University Press, 1978.

Kurzman, Dan. *Genesis, 1948.* London: Vallentine, Mitchell, 1970.

Laska, Vera, ed. *Women in the Resistance and the Holocaust.* Westport, Conn.: Greenwood Press, 1983.

Lernoux, Penny. *Cry of the People.* Harmondsworth: Penguin, 1980.

Maclear, Michael. *The Ten Thousand Day War: Vietnam 1945-75.* London: Methuen, 1981.

Malraux, André. *Anti-Memoirs.* Translated by Terence Kilmartin. New York: Holt, Rinehart and Winston, 1968.

Marron, Juddy. Letter to the *Vet Reporter,* 1 July 1982.

Marshall, Bruce. *The White Rabbit.* London: Evans Brothers, 1952.

Masson, Madeline. *Christine.* London: Hamish Hamilton, 1975.

Meshad, Shad. Interview in *The New York Times,* 23 March 1981.

Minerva: Quarterly Report on Women and the Military (Arlington). Edited by Linda Grant de Pauw.

Mountfield, David. *The Partisans.* London: Hamlyn Pub. Group, 1979.

Myles, Bruce. *Night Witches: The Untold Story of Soviet Women in Combat.* Novato, Calif.: Presidio Press, 1981.

North, Liisa. *Bitter Grounds: Roots of Revolt in El Salvador.* Toronto: Between the Lines, 1981.

Randall, Margaret. *Sandino's Daughters: Testimonies of Nicaraguan Women in Struggle.* London: Zed Press, 1981.

Schoenbrun, David. *Soldiers of the Night: The Story of the French Resistance.* New York: Dutton, 1980.

Smith, Denis M. *Mussolini.* New York: Vintage Books, 1983.

Tillion, Germaine. *Ravensbruck: An Eyewitness Account of Women's Concentration Camp.* Translated by Gerald Satterwhite. New York: Doubleday, 1975.

Tompkins, Peter. *A Spy in Rome.* New York: Simon & Schuster, 1962.

Valery, Anne. *The Passing Out Parade: A Play.* London: Samuel French, 1980.

Van Devanter, Lynda. *Home before Morning: The Story of an Army Nurse in Vietnam.* New York: Beaufort Books, 1983.

Willenz, June. *Women Veterans: America's Forgotten Heroines.* New York: Continuum, 1983.

Women in Combat and as Military Leaders. Unpublished. Staff Report Branch of the Center of Military History. U.S. Army, 1978.

Zawodny, J.K. *Nothing but Honour: The Story of the Warsaw Uprising.* Stanford, Calif.: Hoover Institute Press, 1978.

RESOURCES

Amnesty International, Toronto headquarters

Association of Women for Peace, Leningrad

Association of Women of El Salvador (AMES)

Development Education Centre (DEC), Toronto

Imperial War Museum, London

Israeli Consulate, Toronto

Kinneret Foundation, Washington, D.C.

Latin America Working Group (LAWG), Toronto

Ministry of Defence, London

Ministry of Defence, Paris

National Association of Italian Partisans (ANPI), Rome

Novosti Press Agency, Moscow

Pentagon Center of Military History, Washington, D.C.

Polish Home Army Veterans' Association, Toronto

Robarts Library, University of Toronto

Royal Canadian Legion, Ottawa

Soviet Press Office, Soviet Embassy, Ottawa

Toronto Metropolitan Reference Library

Vietnam Nurses' Project, Minneapolis, Minn.

Vietnam Veterans of America, Lansing, Michigan

Vietnam Veterans of America, Women's Steering Committee, Washington, D.C.

The Vietnamese Women's Union, Hanoi

War Veterans' Committee, Moscow